DESIGN ASPECTS OF USED LUBRICATING OIL RE-REFINING

DESIGN ASPECTS OF USED LUBRICATING OIL RE-REFINING

FIRAS AWAJA

RMIT University, Melbourne, Australia

DUMITRU PAVEL

University of Notre Dame, Notre Dame, IN, USA

ELSEVIER

AMSTERDAM • BOSTON • HEIDELBERG • LONDON • NEW YORK • OXFORD
PARIS • SAN DIEGO • SAN FRANCISCO • SINGAPORE • SYDNEY • TOKYO

Elsevier
Radarweg 29, PO Box 211, 1000 AE Amsterdam, The Netherlands
The Boulevard, Langford Lane, Kidlington, Oxford OX5 1GB, UK

First edition 2006

Library of Congress Cataloging-in-Publication Data
A catalog record for this book is available from the Library of Congress

British Library Cataloguing in Publication Data
A catalogue record for this book is available from the British Library

ISBN-13: 978-0-444-52228-3
ISBN-10: 0-444-52228-X

For information on all Elsevier publications
visit our website at books.elsevier.com

Printed and bound in The Netherlands

06 07 08 09 10 10 9 8 7 6 5 4 3 2 1

To our families for believing in us

Contents

Preface

Huge amounts of used lubricating oils from automotive sources are disposed of as a harmful waste into the environment. For this reason, means to recover and reuse these wastes need to be found. Valuable lubricant could be recovered from disposed oil. More importantly, oil may be the most important factor to influence the shape of society in the 21st century. The cost and availability of oil and its products will significantly impact our quality of life, the stability of our environment, the health of national economies and even the relationships between nations.

The purpose of this book is to present a feasible and comprehensive technology for recycling used lubricating oil. Furthermore, this book demonstrates experimental information on the subject of lubricating oils behavior under the process conditions of distillation and solvent extraction. The book presents essential experimental results for process designers and engineers for establishing a complete process design. The detailed information can be scaled up directly into a pilot plant towards a commercial size production plant. The book also introduces background information on the lubrication oil recycling industry, outlining the major manufacturers and detailing their processes. The text is also useful for engineering and industrial chemistry students in that its methodologies could be useful in their project design and enhance their analytical knowledge of a chemical process with respect to productivity and efficiency.

The study reported herein involved the re-refining of used lubricating oil taken from different automobiles. The re-refining process consists of dehydration, solvent extraction, solvent stripping, and vacuum distillation. The lubricating oil processing conditions and behavior were examined for each process step. The re-refined oil properties were compared with virgin and used oils and their distillation behaviors were compared.

By vacuum distillation, highly pure lubricant cut with approximately zero ash content was obtained using solvent treated oil with an optimum reflex ratio of 4 to 1 and pressure of 4 mBar. Additives were added to the final product to improve viscosity and pour point properties and to compare with the commercial virgin lubricating oil.

Acknowledgment

The authors would like to express their gratitude to the many people that helped in different ways towards the completion of this book. The first author thanks Prof. G.A. Rassoul for insightful discussions during the period of the experimental work and the lab technicians and chemists who helped in the material analysis and characterization.

Chapter 1

Introduction

The processes of re-refining of used lubricating oils depend greatly on the nature of the oil base stock and on the nature and amount of contaminants in the lubricant resulting from operations. The contaminants are introduced either from the surrounding air and from the engine, which are called extraneous contaminants, or from the products of oil deterioration. These contaminants must be totally removed in order to reuse the lubricating oil. Processes for recovery of used oil vary from simple settling or filtration to solvent extraction or vacuum distillation or a combination of these and other processes.

Preliminary study of Iraqi used lubricating oil did not involve pre-treatment steps before distillation, while international studies show complete processes that involve pre-treatment, vacuum distillation and finishing steps such as clay contacting, hydro treating and blending. Neither the distillation behaviors nor optimum procedures to evaluate the operating conditions were reported by literatures for refining used oil. Iraqi lubricating oil properties differ from western lubricating oil properties; therefore, it was not possible to compare their process operating conditions.

This study aims to investigate a process of solvent extraction of an alcohol–ketone mixture as a pre-treatment step followed by vacuum distillation. The primary step was conducted before the solvent extraction that involves dehydration to remove the water and fuel contaminants from the used oil. The solvent extraction and vacuum distillation steps were used to remove higher molecular weight contaminants. Experiments were conducted on an Iraqi mixed base stock used lubricating oil to produce lubricating cuts free of contaminants. Recovered lubricating cut can be blended with original lubricating cuts from the refinery. Additives may be used to improve the quality of re-refined lubricating oil.

Chapter 2

Background

2.1 Virgin Oil Characteristics

Automotive lubricating oil is made by introducing proper additives to lubricating oil cuts manufactured from paraffin or mixed-base crude oils (Nelson, 1964; Forbes and Neustadter, 1972). The main functions of lubricating oil include reducing friction, carrying away heat, protecting against rust, protecting against wear and removing contaminants from the engine. A typical blend of lubricating oil is shown in Table 2.1. Table 2.2 shows the typical properties of virgin crankcase oil according to the Iraqi and international standards. The additives are used to enhance the natural properties of the lubricating oil and to prevent some undesirable properties. The main additives for lubricating oil are as follows.

2.1.1 Oxidation Inhibitor

An oxidation inhibitor is used to prevent or eliminate the oxidation reactions of lubricating oil. The oxidation reactions conditions are present in the crankcase of the engine. These reaction conditions include high temperature, the presence of air, and the presence of metal in the engine. The metals act as catalyst for the oxidation reactions. Furthermore, the products of oil oxidations act as the catalyst to accelerate further the rate of oxidation. Oil oxidation usually occurs in a series of steps in which substances are formed such as petroleum hydrocarbons, aldehydes, organic acids, resinous substances, varnish-like materials, lacquers, carbonaceous material, and coke (Guthrie, 1960).

The amount of lubricating oil that is lost in the oxidation reactions is small. However, the products of oil oxidation interfere with lubrication by forming sludge and corrosive byproducts to the extent that the oil must be either replaced or purified. A variety of organic chemical compounds have been used as oxidation inhibitors such as sulfur compounds, phosphor compounds, sulfur/phosphor compounds, amines and phenolic derivatives (Gruse and Stevens, 1942; Braithwaite, 1967).

Table 2.1. Typical blend of lubricating oil

Gradient	Percentage
Base oil	86
Viscosity index improvers (polyisobutylene, polymethacrylate)	5
Oxidation inhibitor (zinc dialkyl, dithio phosphate)	1
Detergent (barium and calcium sulphonates or phenates)	4
Multi-functional additives (dispersant, pour point depressant)	4

Table 2.2. Typical properties of different grades of crankcase oils

(a)* Iraqi standards

Grade	Viscosity (cSt) 100 °C	Pour point (°C)	Flash point (°C)	V.I.	Sp gr
10 W	4.5–6	−27	210	95	0.875
20–20 W	6.5–8.5	−21	220	95	0.885
30	9.5–12.5	−18	228	95	0.890
40	13–16.5	−12	236	95	0.895
50	17–21	−12	240	95	0.898

(b)† International standards

Grade	Viscosity (cSt) 40 °C	Viscosity (cSt) 100 °C	Pour point (°C)	Flash point (°C)	V.I.
Sw-40	55.8	9.5	−38	218	155
10W-30	61.9	9.6	−35	224	137
10W-40	96.7	14.4	−35	218	154
10 W	37.1	6.1	−33	205	108
20 W	68.8	9.0	−27	224	104
30	111.0	12.1	−21	230	98
40	114.8	14.5	−18	240	98
50	210.7	18.6	−12	245	98

*Iraqi ministry of oil publication, 1993.
†McKetta, 1986.

2.1.2 Detergent and Dispersant Additives

Detergent and dispersant are used to keep particulate matter in suspension or dispersion. Particulate consists of fuel soot, resins and oil oxidation products. By remaining suspended, these products will neither adhere to metal surfaces as varnish or lacquers, nor settle out in the engine sump as sludge. Detergent additive produces also anti-oxidant and anti-corrosive properties, while the dispersant, being polymeric in nature, may also improve the viscosity index (Braithwaite, 1967).

2.1.3 Viscosity Index Improvers

The viscosity index is very important because it indicates the suitability of an oil to lubricate properly at the elevated temperature of the engine. At the same time, lubricating oil should be moderately viscous for operating during starting. Polyisobutylene and polymethacrylate are used as viscosity index improvers for lubricating oil (Nelson, 1964).

2.1.4 Pour Point Depressants

Pour point depressants are used to lower the pour point temperature of the lubricating oil beyond the starting temperature of the engine. The pour point depressants commercially employed are polyalkylnaphthalene, polyalkyphenyl esters or polyalkylmethacrylate (Braithwaite, 1967).

2.2 Used Lubricating Oil Characteristics

The automotive lubricating oil loses its effectiveness during operation due to the presence of certain types of contaminants. These contaminants can be divided into:

2.2.1 Extraneous Contaminants

Extraneous contaminants are introduced from the surrounding air and by metallic particles from the engine. Contaminants from the air are dust, dirt, and moisture. Air itself may be considered as a contaminant since it might cause foaming of the oil. The contaminants from the engine are:

(1) Metallic particles resulting from wear of the engine,
(2) Carbonaceous particles due to incomplete fuel combustion,
(3) Metallic oxides present as corrosion products of metals,
(4) Water from leakage of the cooling system,
(5) Water as a product of fuel combustion and
(6) Fuel or fuel additives or their byproducts, which might enter the crankcase of engines (Guthrie, 1960).

2.2.2 Products of Oil Deterioration

Many products are formed during oil deterioration. Some of these important products are:

(1) Sludge: a mixture of oil, water, dust, dirt, and carbon particles that results from the incomplete combustion of the fuels. Sludge may deposit on various parts of the engine or remain in colloidal dispersion in the oil.
(2) Lacquer: a hard or gummy substance that deposits on engine parts as a result of subjecting sludge in the oil to high temperature operation.
(3) Oil-soluble products: the result of oil oxidation products that remain in the oil and cannot be filtered out and deposit on the engine parts. The quantity and distribution of engine deposits vary widely depending on the conditions at which the engine is operated. At low crankcase temperatures, carbonaceous deposits originate mainly from incomplete combustion products of the fuel and not from the lubricating oil. While, at high temperature, the increase in lacquer and sludge deposits may be caused by the lubricating oil (Kalichevsky and Kobe, 1956; Haycock et al., 1988).

2.3 Effects of Oil Contaminants

The lubricating oil properties are affected by any contaminants that may occur during motor operation. The effects of the contaminants are as follows:

(1) Water: even in small amounts, water causes rusting of iron or steel. The water also results in forming water sludge (emulsions), which may clog oil passages, pump valves and other oil handling equipment. Water also contributes to foaming problems.

(2) Solid particles of dirt, dust, grit and metallic fragments, which were circulated by the lubricant: these contaminants cause excessive wear, scoring of bearing surface, and possible failure due to seizing of metal fatigue.
(3) Sludge and lacquers: the sludge deposits clog small oil passages and clearances. Lacquers or varnish cause sticking of valves, and resist against the continuous operation of oil pump.
(4) Liquid contaminants such as unburned fuel from engines dilute lubricating oil and possibly reduce their viscosity beyond a safe load. Contaminants of the lubricant with heavier oil increase viscosity and interfere with the oil circulation. This affects the lubricating valve and heat transfer capacity (Guthrie, 1960).

2.4 The Hydrocarbon Composition of Used Oil

The hydrocarbon composition of new or used automotive lubricating oil sludge consists primarily of saturated compounds such as linear and branched chain, paraffins, which have at least twice as many naphthenes. Aromatics generally comprise about 10 to 15 weight percent of the hydrocarbon base material (Hess, 1979; Wang et al., 2000).

Cotton et al. (1977) analyzed 30 different used oils. They found that the composition of the used oil consists of four major groups based on the 30 oil samples analysis, which are shown to be for average values of 76.7% saturates, 13.2% monoaromatics, 3.7% diaromatics and 6.5% polyaromatics-polar material. They used an analytical procedure that includes silica–alumina gel adsorption chromatography and mass spectrum analysis.

Mohammed et al. (1984) analyzed the composition of two samples of Iraqi base stock used lubricating oil distillate fractions (350 °C to 550 °C) by using elution adsorption chromatography (EAC). The composition result for olefins, saturates and aromatics were 5.5%, 73.3% and 21.5% for sample (1) and 7.4%, 73.7% and 18.9% for sample (2) respectively.

2.5 Physical and Chemical Tests of Used Lubricating Oil

Standard chemical and physical tests were used to evaluate the nature and the extent of the contaminants in the used automotive oils. These tests involve the following measurements:

(1) Viscosity: viscosity testing can indicate the presence of contamination in used lubricating oil. The oxidation and polymerization products that were dissolved and suspended in the oil cause the increase of oil viscosity. While a decrease in the viscosity of lubricating oil indicates the fuel contamination.
(2) Pour point: pour point is the lowest temperature at which the oil will flow. Low pour point indicates good lubricating oil.
(3) Flash point: flash point is the lowest temperature at which the vapors in air will burn momentarily if ignited by flame or spark. A decrease in flash point indicates contamination by dilution of lubricating oils with unburned fuel. Increasing of flash point indicates evaporation of the light components from the lubricating oil.
(4) Acidity or neutralization number: this is a measure of the amount of alkali required to neutralize one gram of the oil. An increase in acid number is due to oxidation of lubricating oil.

(5) Ash content: the remaining solid ash, when the oil is completely burned, is a measure of oil purity.
(6) Carbon or coking test: this evaluates the solid residue obtained when the oil is heated to complete vaporization and it refers to the amount of deposit formed.
(7) Water content: this test is done by distillation and indicates the amount of water emulsified in the oil.
(8) Fuel contaminants: this test indicates the amount of fuel diluting in the lubricating oil during automotive operation (Hobson, 1955; McKetta, 1986).

In the following tables, the tests of selected used lubricating oil are summarized. Table 2.3 indicates that the first base oil (Oklahoma city), which has similar crude base characteristics to Kirkuk crude (Nelson, 1964), has relatively high specific gravity, viscosity, flash point and water content. These indicate that this oil has a high percentage of heavy component and water. Furthermore, it has high ash content that leads to high metal content. The second base (Bartlesville, Oklahoma) has a lower specific gravity due to the smaller water content. Nevertheless, it also has a high ash content and high viscosity. The Iraqi sample number 1 has moderate ash content. It also has a very low flash point that indicates a high percentage of fuel dilution. The Iraqi samples have higher sulfur content than other base used oils due to the high sulfur content in the original Iraqi crude oil (Kirkuk crude). The fifth base oil (M.B.T., NY) has low flash point due to the same reason as Iraqi sample number 1. The sixth base oil (NY, USA) has a relatively high carbon residue that indicates higher percentage of solid residue after distillation (Bethea et al., 1973; Cotton et al., 1977; Mohammed et al., 1984).

The elements in the lubricating oil that come from the additives and which are shown in Table 2.4 are Ba, Ca, Mg, P and Zn. The Cu and Fe elements come from the wear of the engine, while the Pb element comes from the wear of the engine and from the fuel additives (Bethea et al., 1973; Cotton et al., 1977; Mohammed et al., 1984).

Table 2.3. Standard tests of different base used oil

Property	Oklahoma (USA)	Bartlesville (USA)	Sample 1 (Iraq)	Sample 2 (Iraq)	Mobil Co. (USA)	NY (USA)
Sp gr	0.929	0.908	0.902	–	0.906	0.908
Viscosity (cSt) 40 °C	92	90	–	41.11	59	68
Viscosity (cSt) 100 °C	13	12.5	–	6.21	9.25	10.3
Water (wt.%)	13.7	0.4	1.2	–	4.4	0
Ash (wt.%)	1.51	2.2	0.781	–	1.81	
Flash point (°C)	207.7	185	110	–	101.7	182.2
Pour point (°C)	−34	−40	–	–	–	−37.2
Carbon residue (wt.%)	3.25	2.18	4.25	–	–	5.2
Sulfur (wt%)	.34	0.47	1.45	1.25	0.34	0.32
V.I.	154	143	–	–	166	141
Fuel (wt.%)	3	1.6	–	–	–	–

Table 2.4. Metal content analysis of different used oils

Element ppm	Oklahoma (USA)	Bartlesville (USA)	Iraq	Mobil Co. (USA)
Ba	204	59	295	600
Ca	1295	2670	1147	1700
Mg	312	338	–	–
Na	79	69	–	–
P	1074	1341	–	900
Zn	1151	1574	590	800
Al	15	25	–	–
Cr	12	14	–	–
Cu	31	16	15	–
Fe	212	342	107	360
K	16	29	–	–
Mn	3	4	–	–
Ni	1	1	–	–
Pb	8450	11165	524	11100
Si	15	14	–	–
Sn	3	14	–	–
V	0	0	–	–

2.6 Used Oil Recovery Processes

There are a large number of physical and chemical processes for reclamation, re-refining and reprocessing of used lubricating oil (Whisman et al., 1978; Reis and Jeronimo, 1990; Omar et al., 1998; Andreev and Tolmachev, 2002; Kim et al., 2003; Ostrikov and Prokhorenkov, 2003; Bhaskar et al., 2004). The earliest process known was the acid–clay treatment process. This treatment involves mixing the used oil with 93 to 98% sulfuric acid. The sulfuric acid acts as an extraction medium for the removal of asphaltenes, unsaturates, dirt, additives, color bodies and other material from the used oil. The acid treated oil will be then mixed with clay (e.g., fuller earth), filtered and reused (Snow and Delaney, 1977; Hess, 1979).

Problems arising from acid treatment include environmental problems associated with the disposal of acid sludge and spent earth, low product yield (45–65%) and incomplete removal of metals, especially lead (Snow and Delaney, 1977). An alternate process using dehydration, distillation and hydro finishing was developed by the national oil recovery corporation (NORCO, USA) in 1969. The feed oil passes through a flash furnace and a tower to separate water and gasoline fractions. The oil is then heated up to 360–370 °C and passed to vacuum fractionators operating at 400 °C and 34 mBar. The column separates the oil into light and heavy oil products (Snow and Delaney, 1977).

Berry (1981) developed a process using a series of vacuum distillation steps involving equipment of a unique design that minimize coking during operation. The main problems encountered in these processes are plugging of the lines, fractionators and furnace tubes due to formation of a resinous material that fouls the equipment. The process did not improve the dark color of the product (Quang et al., 1976; Snow and Delaney, 1977). All modern technologies have included a physical or chemical pre-treatment step in order to avoid or eliminate these problems. The major important pre-treatment processes are:

(1) The process of Leybold–Heraeus of Germany, that includes sodium pre-treatment in which the sodium particles sizes from 5 to 15 micron are dispersed into the dry oil in amounts less than 1%. The contaminants in the oil are transformed into compounds that cannot be distilled such as salts or polymer (Erdweg, 1978; Hess, 1979).
(2) The process of the Institut Francaise du Petrol (IFP) uses propane extraction for separation of heavy ends. This process is similar to that commonly used in refineries to separate asphaltenes for preparation of catalytic cracking feed or new heavy neutral base lubricating oils (McKeagan, 1992).
(3) Whisman et al. (1978) of the U.S. Department of Energy (Bartlesville Energy Research Center, BERC) described a method of treating the dehydrated used oil with a mixture of one part 2-propanol, two parts of 1-butanol and one part of butanone. The solvent to oil ratio was 3 to 1 by volume. By this treatment the alcohol–ketone mixture will reduce coking and fouling problems during distillation (Whisman et al., 1978; Hess, 1979).

The basic steps in any modern used oil re-refining technologies are:

(1) Dehydration of the used oil to remove gasoline and water fractions that result from the automobile engine operation.
(2) Chemical or physical pre-treatment of dehydrated oil.
(3) Fractionation of pre-treated used oil.
(4) Finishing steps needed to improve color, odor, and viscosity index including either clay contacting or hydro treating or blending (Skala et al., 1991). The major differences encountered between modern re-refining technologies were due to pre-treatment and finishing steps.

Table 2.5 Summarizes the important major technologies practiced by major re-refining companies with their advantages and disadvantages (Quang et al., 1976; Whisman et al., 1978; Havemann, 1978; Erdweg, 1978; Pauley, 1978; Hess, 1979; Gruber, 1992; Patel, 1994).

2.6.1 Dehydration of Used Oil

The water and gasoline component in the used lubricating oil must be distilled off because their presence changes the lubricating oil properties. This may affect the validity of any pre-treatment step especially the solvent treatment (Whisman et al., 1978). The dehydration conditions (Table 2.6) depend on the amount of water and gasoline fractions in the used oil.

2.6.2 Solvent Treatment

Many researchers investigated solvent treatment/extraction of used lubricating oil (Snow and Delaney, 1977; Whisman et al., 1978; Reis and Jeronimo, 1988, 1990; Martins, 1997; Elbashir et al., 2002; Hamad et al., 2005). Whisman et al. (1978) introduced a mixture of composite solvents for treating used oil using 2-propanol, 1-butanol and butanone. Their

Table 2.5. Important re-refining technologies

Technology	Pre-treatment	Finishing	Advantages	Disadvantages	Reference
VISCOLUBE (Italy)	Propane solvent extraction	Clay contacting	High quality and yield	Expensive and fire hazards	Viscolube, 2005
Institute Francias du Petrol (IFP) (France)	Propane solvent extraction	Hydro treating or clay contacting	High quality and yield	Expensive and fire hazards	IFP, 2005
Bartlesville energy research center (USA)	Alcohol–keton solvent extraction	Hydro treating or clay contacting	High quality and yield		Bartlesville, 2005
Kinetics tech. Int.(KTI)	Chemical (not specified)	Hydro treating	High quality and yield and clean waste	High capital cost	KTI, 2005
Chemical Eng. Partners (CEP, USA)					CEP, 2005
Safety clean (USA)					
Evergreen (USA)					
Mohawk (Canada)					
Recyclon leybold (Germany)	Sodium	Clay contacting	Low capital cost	Environmental problems	Evergreen, 2005

Table 2.6. Some dehydration conditions as reported by different researchers

Temperature (°C)	Pressure (mm Hg)	Reference
173	2	Bishop and Arlidge, 1978
148	760	Gruber, 1992
148	10	Whisman et al., 1978
250	760	Scott and Hargreaves, 1991
218	760	Berry, 1981

investigations showed that the two alcohols make a binary system that was reasonably effective. However, best results were obtained at a solvent to oil ratio of 8 to 1, which was considered unfeasible. They evaluated various ratios of the ternary systems and noticed that the 2-propanol rich systems produced poorly separated sludges. While butanone rich systems produced good sludge separation but it seemed to re-dissolve segregated ash-forming materials. Therefore they chose a solvent system of (1:2:1) of (2-propanol, 1-butanol, butanone) with a solvent to oil ratio of 3 volumes to 1. Snow and Delaney (1977) investigated the solvent extraction systems of used oils. The solvent to oil ratio was either 4 to 1 or 8 to 1. The solvents studied were 2-propanol\1-butanol mixture, methyl isobutanone, butanone and acetone. They concluded that with respect to the reduction of resin, the best pre-treatments were the propanol\butanol solvent and acetone solvent.

Reis and Jeronimo (1988) studied the performance of ketones and alcohols, which are miscible with base oil at room temperature and their flocculating action on oil contamination. The results of their study indicate that the flocculating action of polar solvents in used oils is an anti-solvent effect exerted on some nonpolar macromolecules. They showed that the difference in solubility parameters ($S_1 - S_2$) between the solvent and a typical polyolefin (polyisobutylene), which is considered as a contaminant in the used oil, was used as a preliminary criterion to select the components of composite solvents. They noticed in some cases the polar solvent induces the formation of an electrically stabilized dispersion and recommended the addition of potassium hydroxide in alcoholic solution to easily destabilize the dispersion and increase sludge removal from used oils. They called a solvent an extraction–flocculation solvent if it has the following properties:

(1) Miscibility with the base oil contained in the used oil being processed.
(2) It must exclude from the solution the additives and the dispersed particles (or part of them) allowing their aggregation to particle size big enough to be separated from the liquid by sedimentation.

The authors noticed that the ketone and alcohols having four carbon atoms or greater are extraction–flocculation solvents. They showed that the flocculating actions for alcohols decrease from 1-butanol to 1-hexanol in system of experiments containing (1-butanol, *sec.* butyl alcohol, 1-pentanol and 1-hexanol). For ketones, butanone exhibits a more active flocculating action than methyl *n*-propylketone, which in turn is more effective than methyl isobutyl ketone. They showed that when there is no segregation of polymeric molecules from solution the black particles remain in stable dispersion. But when flocculation occurs, macromolecules and carbonaceous particles co-flocculate and settle together. In a separate study, Reis and Jeronimo (1990) recommended that the capability of polar solvents to segregate sludge from used oils is closely related to their solubility parameters. The polarity itself, as measured by dipole moment, is not correlated with that capability of flocculation.

Referring to the solubility parameter difference values ($S_1 - S_2$), 2-propanol would be a better oil-flocculating agent than 1-butanol with ethanol even better at flocculating than 2-propanol. Propanone would be more efficient than butanone; however, none of these alcohols and ketones having less than four carbon atoms are miscible with the base oil operator. The authors reported that by proper selection of components and compositions, formulation having a more favorable balance between miscibility with base oil and flocculation action could be obtained. Van Dyk et al. (1985) best explain the degree of solvency and solubility parameter difference for 34 different polar solvents with polymers.

Elbashir et al. (2002) used a mixture of 2-propanol, 1-butanol and methyl-ethyl ketone (MEK) for used lubricating oil solvent extraction process. They investigated the parameters of solvent type, solvent to oil ratio and extraction temperature. They found out that MEK shows best performance among the three solvents and that the anti-solvency energy, which is a criterion drawn from the difference of solubility parameters of the solvents and oil, depends on the solvent to oil ratio.

2.6.3 Vacuum Distillation

2.6.3.1 *Used oil without pre-treatment.* Distillation as a tool for re-refining used crankcase oil has been used previously by (Brinkman and Dickson, 1995). Whisman et al. (1974) reported experiments on the distillation of used crankcase oil by a batch still with 5 to 1 reflux ratio in a column 30 mm by 122 cm of about 15 theoretical plates. The distillation was conducted at 10 mm Hg pressure. The still pot temperature was 385 °C (equivalent temperature at atmospheric pressure). Practically no product was obtained. The oil in the distillation pot formed a gelatinous mass. Considerable cracking in used oil was indicated by odor development. Continuous distillation column showed improved results. The column was an Asco 4-inch Rota-film (wiped wall) still designed such that sample residence in the heated zone is only a matter of seconds. The authors showed that it was necessary to adjust temperature and pressure to obtain about 20 to 30% bottoms in order to keep the residue fluid. Even under these conditions, heat lumps and other external heat sources were required to keep the residue fluid viscous. Sample was obtained for complete quality testing by operating at 225 °C and 20 mm Hg pressure. Distillation of 1175 g of used oil required 12 hours. A faster rate might have been possible if the used oil first had been dehydrated and gasoline contaminants removed. They concluded that distillation was effective in removal of inorganic materials as shown by carbon residue, sulfated ash, and pentane insoluble. Flash point of the used oil showed evidence of gasoline dilution, which was removed by distillation. The authors concluded that this part of the sulfur was associated with additive compounds while the reminder was originating in the base stock.

Snow and Delaney (1977) experimental results showed that in the straight-run distillation of the used oil, high oil cut boiling between 150 °C and 270 °C at 6 mBar was obtained. The main oil cut between 270 and 360 °C followed this fraction. The quantity of light oil was not more than 15% (w/w) of the oil charged. They indicate that the most serious problem encountered in vacuum distillation is the formation of a resinous material that fouls the equipment. Many researchers noticed this problem. Whisman et al. (1978), suggested a pre-treatment step to reduce coking and fouling in the subsequent fractional distillation. While Quang et al. (1976) reported that processing difficulties occur when the oil contains some polymeric components that were used as a base for various additives. The fractional distillation concentrates the heavy and unstable components into a few fractions and the heat of distillation causes condensation and degradation. These lead to accelerated fouling in heaters and on trays of the film distillation column. A proprietary process to pre-treat the feedstock, because of the additives in the used oil that cause severe coking and corrosion during the re-refining process, were used by the Evergreen company (Gruber, 1992; Evergreen, 2005).

Mohammed et al. (1984) investigated the vacuum distillation to Iraqi base used lubricating oil; the used oil distillation was performed on a bench scale distillation unit. This unit had a 10-tray oldershaw column. Firstly, the emulsified water was removed from the used oil by heating to about 300 °C at atmospheric pressure, using reflux a ratio of 3. The vacuum distillation line of fractions up to 480 °C was carried out at a pressure of 1–10 mm Hg using a reflux ratio of 3. While the separation above 480 °C was continued without the distillation column at 1 mm Hg.

2.6.3.2 *Used oil with solvent extraction pre-treatment.* Snow and Delaney (1977) showed that the solvent-stripped oils were black in color. The vacuum distillation of these oils resulted in a product containing negligible metals, ash, and a marked reduction in the acid number.

Whisman et al. (1978) showed that distillations of the solvent treated oil after removal of the solvent were performed in a variety of bench-scale and pilot plant systems. Typical results were presented in Table 2.7 which provides the equivalent boiling point range at

Table 2.7. Fraction from solvent-stripped oil

Boiling point (°C)	Yields (vol.%)	Viscosity, cSt (40 °C)
IBP-371	2.1	
371-404	26.4	20.0
407-427	23.6	38.0
427-463	25	
463+	14.3	
Loss	8.6	
Total	100	

atmospheric pressure, yield, and lubricating oil fraction viscosity. The operation conditions of the process plant reported that the vacuum distillation of the treated oil was at 5 mm Hg to minimize cracking and to maximize yield. The bottom temperature was maintained at 316 °C, so there is some cracking occurs. This is not in line with that reported by Nelson (1964), which indicated a cracking temperature of about 370 °C.

Chapter 3

Experimental Methods

3.1 Materials

Virgin oil was purchased from the local market; used oil was collected from different automobiles with different operating conditions. Propanol, butanol and butanone were purchased from Aldrich with 99.7% purity. Industrial grade Hexan solvent was obtained from the local refinery (Al Dora) with 95% purity.

The experimental design is a procedure to design efficient experiments. These experiments will yield meaningful results, a minimum of false starts and avoid unnecessary work (Miller, 1986).

3.2 Solvent Extraction Experimental Design

Investigations of proper solvent composition require an experimental design to cover the variable operating ranges with a minimum number of experiments and for detecting interactions between them. This investigation is based on three variables; the first variable (x_1) is the 2-propanol solvent volume percentage with an operating range of 10–40%. The second variable (x_2) is the 1-butanol solvent volume percentage with an operating range of 35–65%. The third variable (x_3) is the butanone solvent volume percentage with an operating range of 10–40%. The center point of operating ranges of these three variables was set to be equal to the values of the best composite solvents, which represent maximum response as reported by Whisman et al. (1978). These variables give two response functions, namely, weight percentage of oil recovery and weight percentage of ash reduction.

The central composite design method (Cochran and Cox, 1975) was chosen to represent the experimental design method. Polynomial models were chosen to fit the experiments, due to the advantages of easy fit and good approximation. The number of experiments required (N), according to this method is

$$N = k^2 + 2k + 1, \tag{3.1}$$

where k = number of variables. The first term ($2k$) represents the number of experiments required to fit the linear response surface. While the second and third term ($2k + 1$) are added to fit a quadratic surface and it is needed to create the central composite design. The last term could be more than one, it represents the repetition experiments in the center (Cochran and Cox, 1975). Hence for three variables, 15 experiments are required.

The first step in central composite design method is to set up the relationship between coded x-scales and the real scales. This can be done by using Eq. (3.2):

$$x = \frac{x_{\text{actual}} - x_{\text{center}}}{\frac{x_{\text{center}} - x_{\text{minimum}}}{\sqrt{k}}}. \tag{3.2}$$

The relationship between the coded and real variables values is shown in Table 3.1. The experimental design for the three variables is shown in Table 3.2.

Table 3.1. The real variables with their coded values

Real values	Coded values				
	−1.73	−1	0	1	1.73
x_1 (%)	10	16.34	25	33.66	40
x_2 (%)	35	41.34	50	58.66	65
x_3 (%)	10	16.34	25	33.66	40

Table 3.2. Experimental design

Experiment	x_1	x_2	x_3
1	−1	−1	−1
2	1	−1	−1
3	−1	1	−1
4	1	1	−1
5	−1	−1	1
6	1	−1	1
7	−1	1	1
8	1	1	1
9	−1.73	0	0
10	1.73	0	0
11	0	−1.73	0
12	0	1.73	0
13	0	0	−1.73
14	0	0	1.73
15	0	0	0

It is obvious that the summation of the real values of the three variables will not be equal to 100%. Adjustments of the variables were carried out in order to achieve total summation equal to 100%.

3.3 Experimental Apparatus

3.3.1 The Oldershaw Sieve Plate Column

The mechanism of the oldershaw sieve plate column is based on the fact that a certain amount of liquid is present on each plate and that the ascending vapor is forced to pass through this liquid. The reflux flows down from each plate to the next plate through tubes (down comers). The holes in the sieve plate (diameter 0.75 to 1 mm) are arranged in circles and pass vertically through the plates (Krell, 1982). The column used in distillation experiments is a 6-plate column with a height of 48 cm. Figures 3.1(a, b, c) indicate the oldershaw sieve plate column, the column at operation and sieve plate layout.

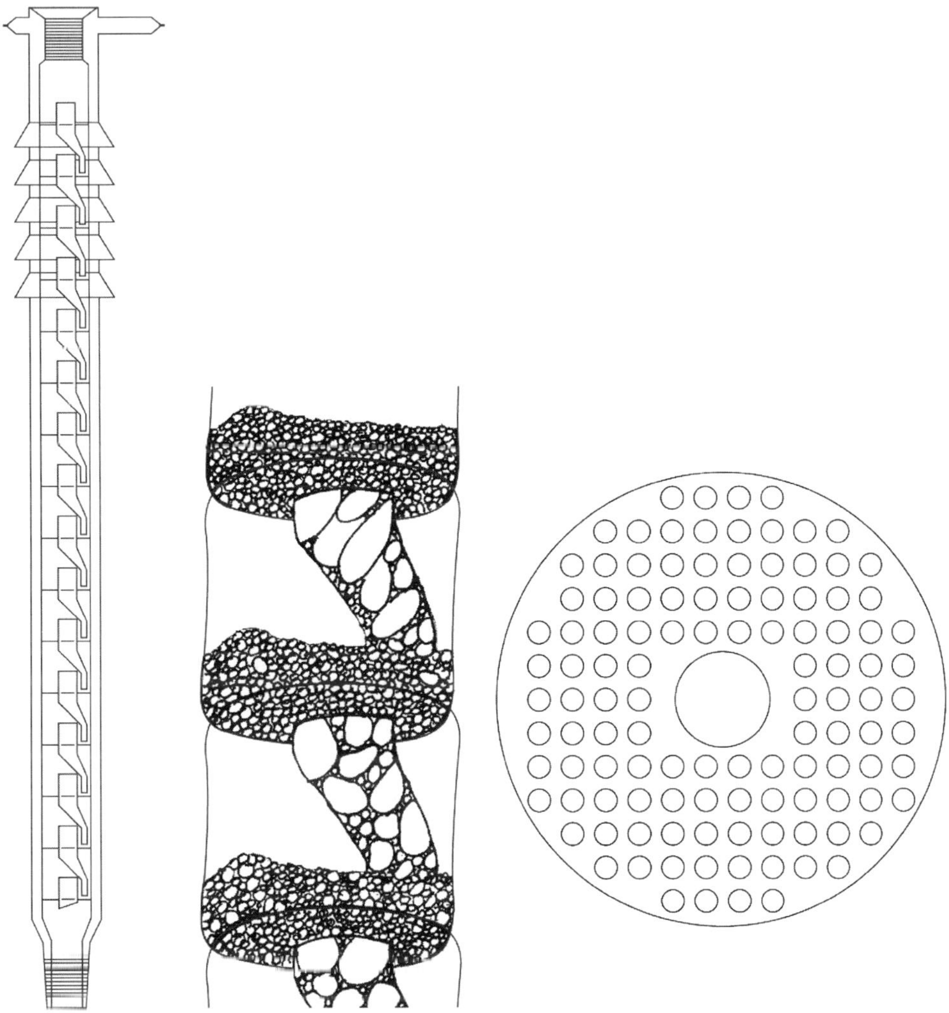

(a) (b) (c)

Fig. 3.1. (a) Oldershaw sieve plate column; (b) the column at operation; (c) sieve plate layout.

3.3.2 The Vacuum Distillation Unit

The vacuum distillation unit consists of the distillation apparatus and the vacuum unit.

3.3.2.1 *The distillation apparatus.* The distillation apparatus consists of heating metal, sieve plate column, divider, condenser and receiver as shown in Fig. 3.2. Heating metal at 600 W was connected to the voltage regulator. The heating metal is capable of holding a still pot of 2 liters volume. A digital temperature indicator with a thermocouple indicates the temperature in the still pot. The inner diameter and cross sectional area of the column is 4.5 cm and 15.9 cm^2 respectively. The column is isolated by a vacuum jacket that makes the outer diameter of the column about 9 cm. The vacuum jacket was plated with a film of insulation material. A vertical glass slide window along the column is available to observe the liquid and vapor behavior in the column. The column was connected from the top with a divider of 27 cm height and about 9 cm outer diameter. The divider has two points for temperature measurement as recorded by thermometers. One of these points is to measure the temperature of the vapor, while the other point is to measure the temperature of liquid condensate. The liquid condensate is collected in the divider by a cone connect with a tube to transfer the liquid condensate to the controlling device that adjusts the reflux ratio. The divider has another tube of 1 cm diameter for the reflux flow. The divider is gathered with condenser that is 22 cm height and 9 cm outer diameter. The condenser has two nozzles for cooling water flow. The condenser was connected from the top with the vacuum line. The liquid product flows from the reflux ratio controller-device through a 2 cm tube to the receiver. The Oldershaw receiver is 100 mL in volume and it is connected with vacuum line through two solenoid valves. A valve was used to connect the receiver with the 1 liter product flask.

3.3.2.2 *The vacuum unit.* The vacuum lines from both the condenser and the receiver are gathered by a T-shaped valve to produce the main vacuum line. A trap was connected to the main vacuum line. The output line from the trap was connected to a one-way valve. The vacuum pressure controller adjusts the plate pressure drop in the distillation system by comparing the vacuum pressure set values and the vacuum pressure values in the trap indicated by the controller sensor. The controller sends its signal to the one-way valve to open by a closed line depending on the vacuum pressure in the trap. The one-way valve was connected to a rotary vacuum pump that produces evacuation to the system. Figures 3.2 and 3.3 show the diagram and view of vacuum distillation of the unit.

3.3.3 Simple Vacuum Distillation Unit

The simple vacuum distillation unit was constructed by removing the sieve plate column from the vacuum distillation unit. Figure 3.4 indicates the simple vacuum distillation.

3.3.4 Simple Atmospheric Distillation Unit

The simple atmospheric distillation unit was constructed by removing the vacuum unit from the simple vacuum distillation unit. Figure 3.5 indicates the simple atmospheric distillation unit.

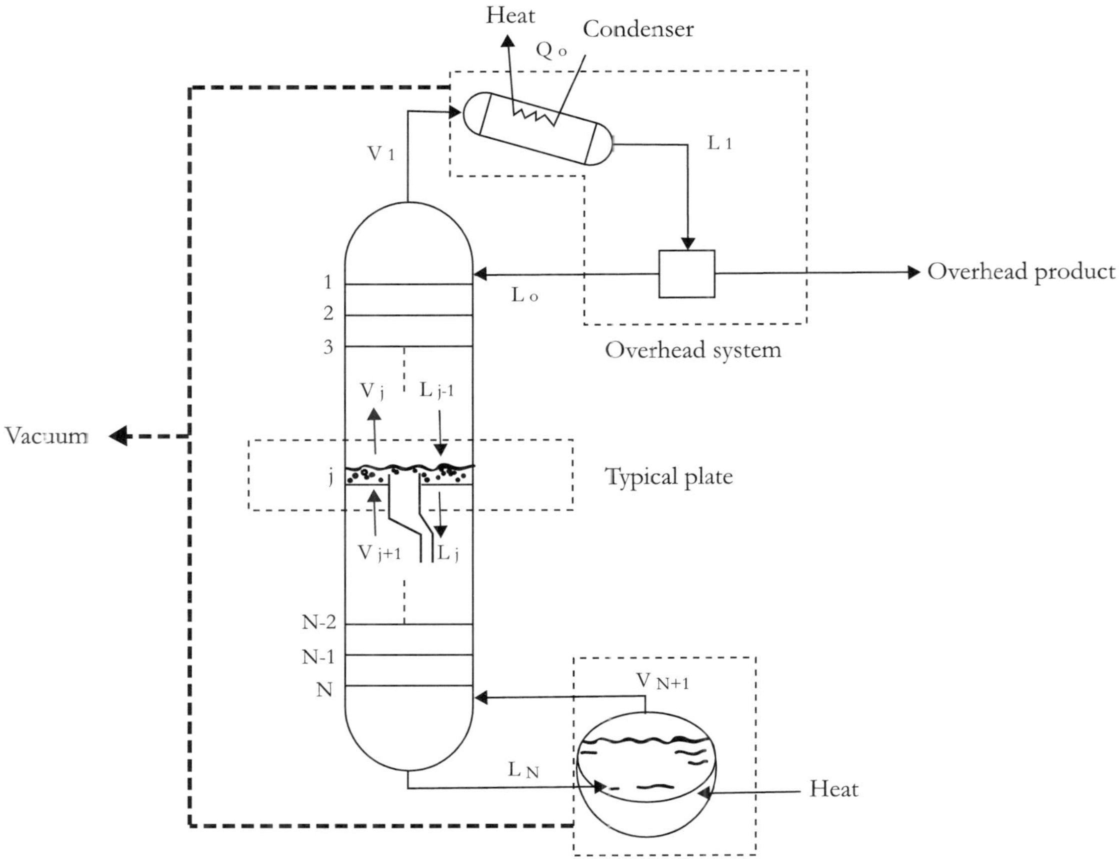

Fig. 3.2. Batch distillation diagram.

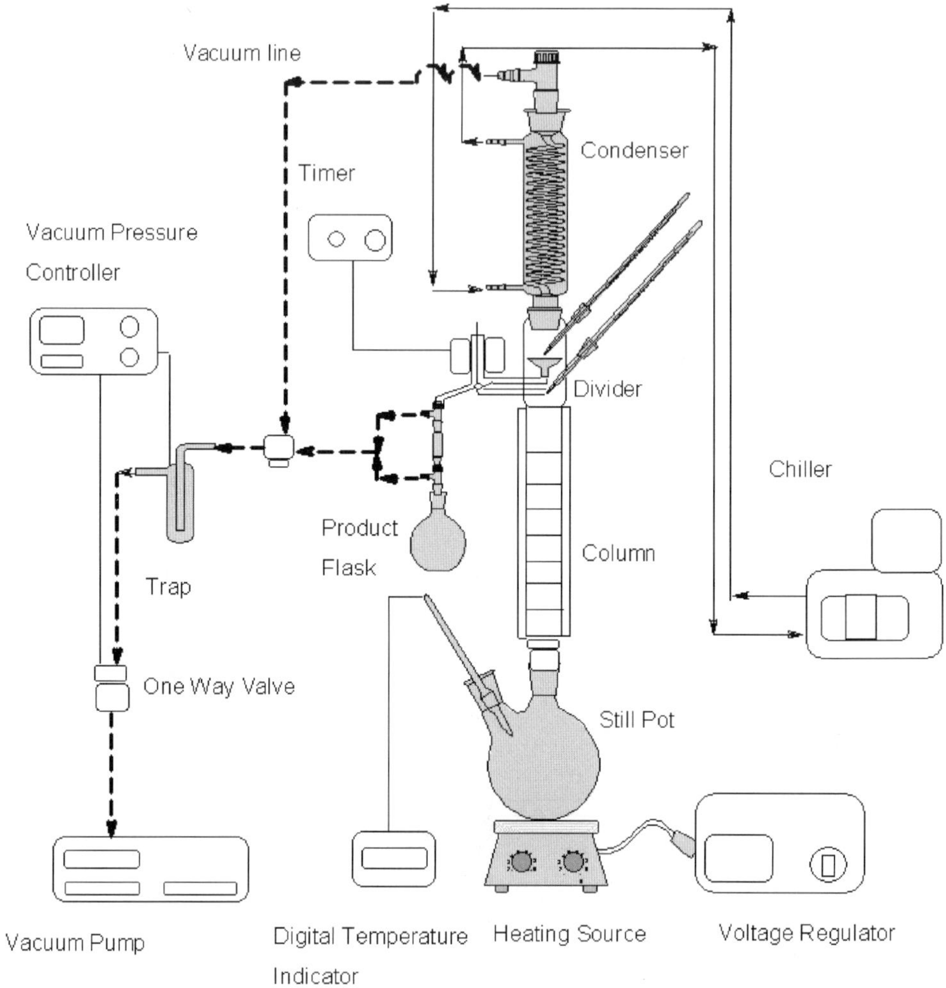

Fig. 3.3. Vacuum distillation unit.

3.4 Experimental Procedures

The used lubricating oils are collected from different automobiles with different operating engine conditions. These oils are mixed together and stored. The used oil mixture was shaken for 5 minutes before any use to insure that the mixture was homogeneous in all parts. All experiments were carried out with used oils taken from this mixture.

3.4.1 Dehydration

The dehydration of used lubricating oil was performed in a simple batch vacuum distillation as shown in Fig. 3.4. Water and gasoline fractions were separated by dehydration of

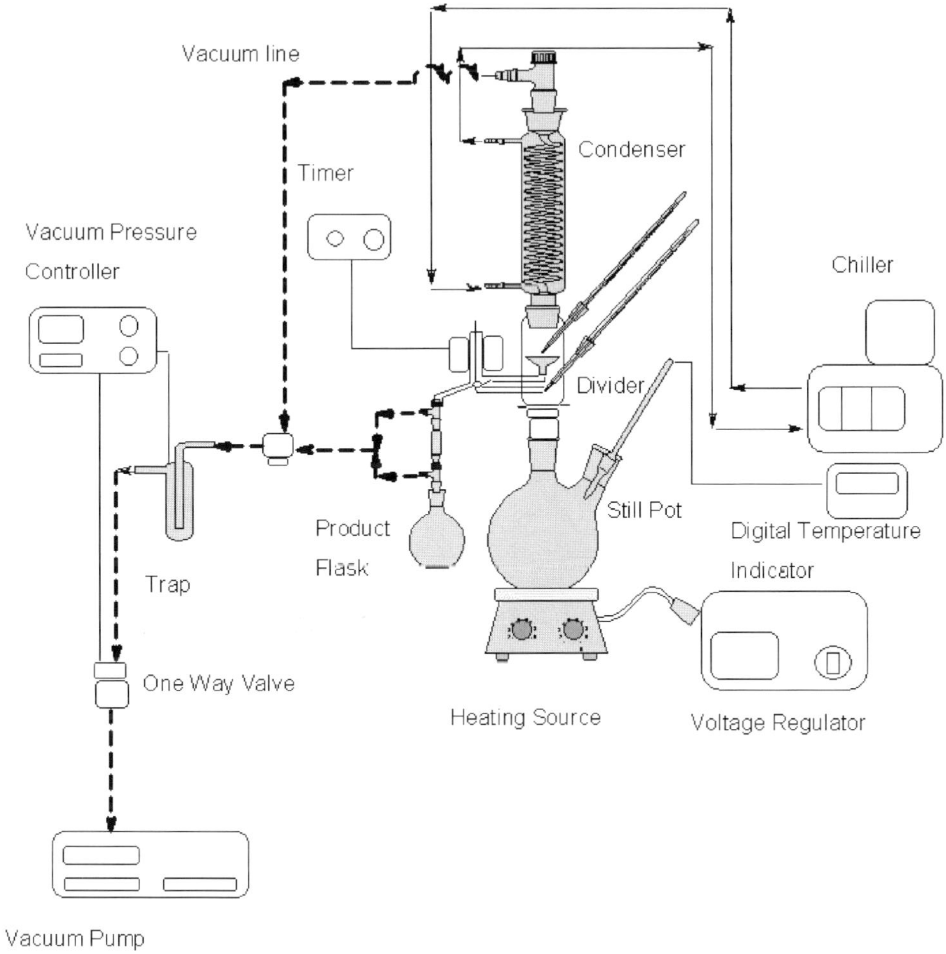

Fig. 3.4. Simple vacuum distillation unit.

used oil. The amount of used oil and heat rate was fixed to be 1.5 L and 600 W respectively. Three different vacuum pressures were used (4, 8 and 12 mBar). Distillation was carried out until no further distillate was produced. The dehydrated used oil was collected and then used for the next step of solvent extraction.

3.4.2 Solvent Extraction

3.4.2.1 *Optimum solvent to oil ratio.* The dehydrated used oil was prepared in amount of 50 mL (about 45 g) for each experiment. The solvent composition was fixed to one part 2-propanol, two parts 1-butanol and one part butanone as reported by Whisman et al. (1978), the investigated solvent to oil ratios were (2, 3, 4, 5 and 6). Solvent to oil ratio less than 2 produced viscous mixture during separation. While when the ratio was higher than

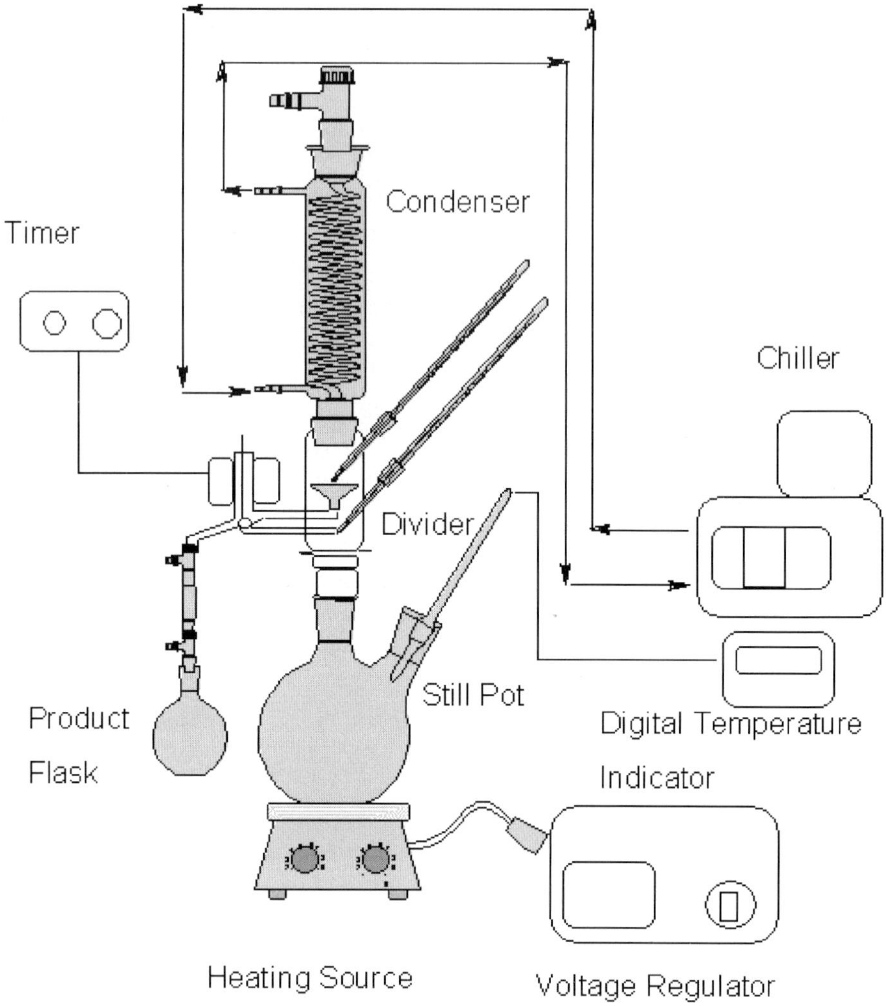

Fig. 3.5. Simple atmospheric distillation unit.

6 to 1, operation was considered economically not feasible. According to these ratios the solvent amounts added were (100, 150, 200, 250 and 300 mL). Adequate mixing of the solvent–oil mixture was obtained by stirring for 15 minutes. The mixture was allowed to settle for 24 hours in order to separate the extract phase (solvents and oil base dissolved) from the raffinate phase (contaminants or sludge). Separation of the two phases was carried out in 1 liter separating funnels. The extract phase was deep red in color and of low viscosity, while the raffinate phase was black and semisolid. After 24 hours the extract phase was separated and turned to another separation funnel and allowed to settle in order to ensure that no contaminants will remain in the solution. If any contaminants (raffinates) appear after 3 hours, the raffinate was separated and the extract was transferred to another

separation funnel again, otherwise the extract was gathered for next step. This procedure was repeated in all experiments for one or two times. The extract phase was subjected to simple batch atmospheric distillation to recover the solvent from the oil by heating up to 200 °C. The produced oil was weighed and tested for ash content.

3.4.2.2 *Optimum solvents composition.* According to the experimental design method, the solvent composition was prepared, which defines the composition of three solvents, i.e. 2-propanol, 1-butanol and butanone. The ratio of three volumes of solvents to one volume of oil was fixed as reported by Whisman et al. (1978). For each experiment, 120 mL (96.4 g) of solvents were mixed with 40 mL (36.2 g) of dehydrated used lubricating oil. The same procedure reported for optimum solvent to oil ratio was used for mixing, separation, solvent recovery, oil weighing and ash content test.

3.4.2.3 *Preparation of large amounts of solvent treated oil.* Larger amounts of solvent treated oil were needed for vacuum distillation experiments. The same procedure mentioned above was conducted to prepare 34 L of solvent–oil mixture. The mixing time was one hour and the settling time was 28 hours, mixing time increases due to larger amounts of the mixture. The solvents stripping or recovery was conducted by atmospheric and then by simple vacuum distillation. In order to reduce the foaming problems, the distillation heat rate was set to 380 W. The temperature of distillation was 200 °C. The last trace of butanol solvent was removed by vacuum distillation at 40 mBar. This was carried out by the units shown in Figs. 3.4, 3.5, and 3.6.

3.4.3 Vacuum Distillation

The vacuum distillation experiments were carried out by the vacuum distillation apparatus shown in Fig. 3.3. After each experiment, the vacuum distillation apparatus were washed with n-hexane solvent in order to remove any contaminants that accumulated in the column, condenser and vacuum lines. The washing procedure was done by atmospheric distillation of one liter of n-hexane at total reflux and 8 hours operation. The n-hexane washed the contaminants and accumulated them at the bottom of the still pot where they can be removed. After washing, all connections and joints were re-lubricated, and prepared for the next experiment. The vacuum distillation operation variables that were studied were:

(1) Reflux ratio: the investigated reflux ratios for solvent treated oils were (0.5, 2, 3, 4 and 6).
(2) Reflex time base: the reflux time base means the time period that set by the timer to control the reflux ratio. The investigated time base was 0.5 s and 5 s.
(3) Heat rate: a heat rate below 400 W is not applicable to the charge amount of oil larger than 700 mL. While heat rate higher than 600 W would accelerate foaming and cracking problems.
(4) Charge: charge amount below 600 mL resulted very low vapor load and a higher amount larger than 1000 mL led to a very long fractionations time in the described apparatus.

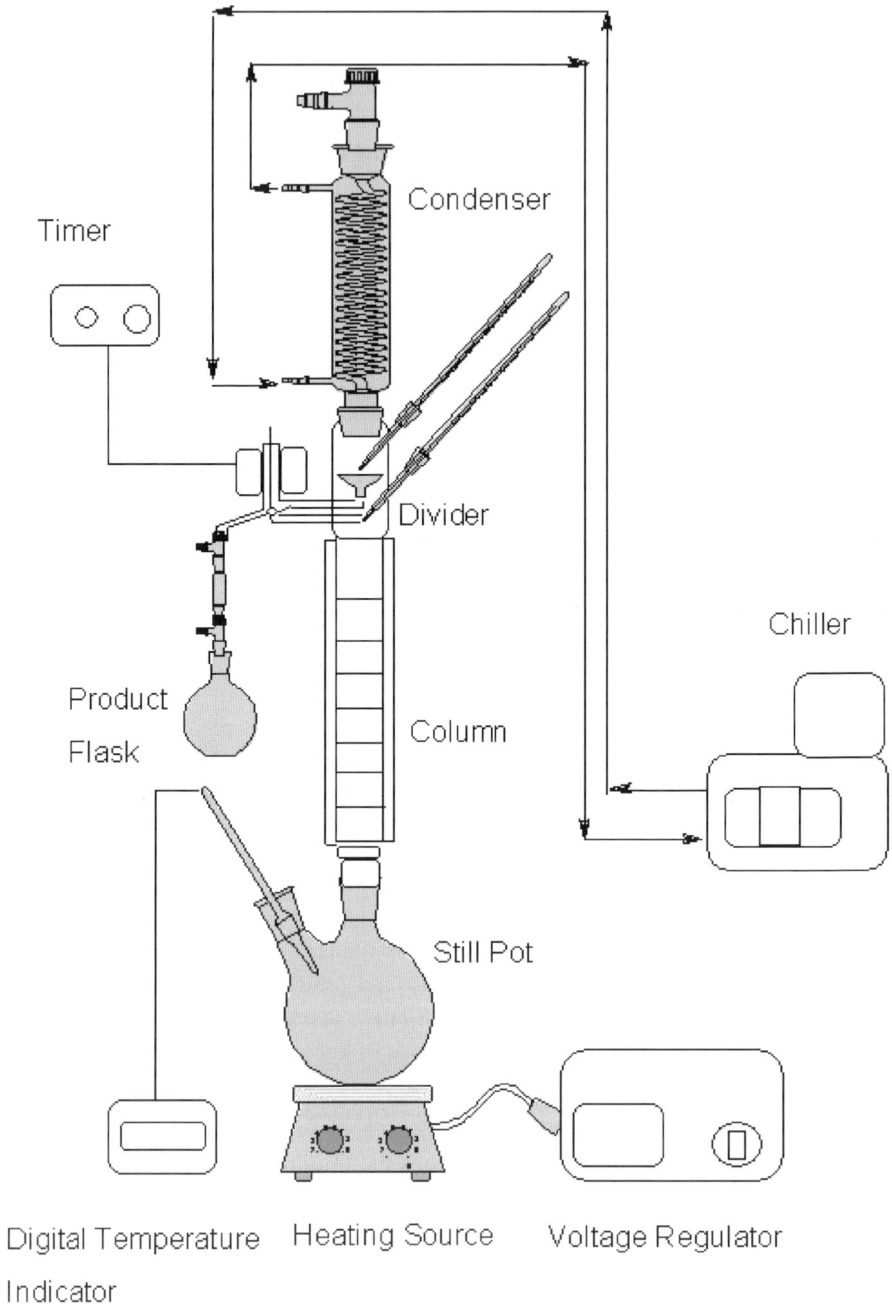

Fig. 3.6. Atmospheric distillation unit.

(5) Vacuum pressure: a vacuum pressure below 4 mBar caused total evacuation of the vapor to the vacuum lines. While vacuum pressure larger than 8 mBar produced a high temperature fractionation, and the temperature rose rapidly far away from the cracking point of 370 °C.

The cooled water temperature in the condenser was fixed to be 15 °C for all experiments. Temperatures lower than 15 °C produced fraction with low viscosity that inhibited flow. The vacuum distillation experiments were performed as follows.

3.4.3.1 *Used lubricating oil.* The behaviors of used lubricating oil fractionation under vacuum distillation were studied by fractionating the following:

(a) A sample of used oil taken from one automobile was fractionated at fixed values for all the above-mentioned variables of reflux ratio, reflux time base, charge, cooling temperature and heat rate.
(b) Samples taken from the used oil mixture were fractionated at different vacuum pressures of 4 and 8 mBar with other variables fixed.
(c) A sample of used oil taken from the used oil mixture was fractionated under an adjusted heat rate to produce final operation temperature not to exceed 370 °C, which is the cracking temperature. The experiment was carried out at fixed variables of reflux ratio, reflux time base pressure, charge, and cooling temperature.

3.4.3.2 *Virgin oil.* The behavior of virgin oil fractionation under vacuum distillation was studied as follows: samples of virgin oil were fractionated at different heat rates (400 and 600 W) and different charges (600 and 700 mL) with fixed other variables of reflux ratio, reflex time base, charge, cooling temperature and pressure.

3.4.3.3 *Solvent treated oil.* After the oil had been treated with solvent. The treated oil was fractionated under vacuum distillation to study the behavior, as follows:

(a) Samples were fractionated at time base of 5 s and at different reflux ratios (0.5, 2, 3, 4 and 6). The fixed variables were pressure, charge, cooling temperature and heat rate.
(b) Samples were fractionated at different pressures (4 and 8 mBar). The fixed variables were reflux ratio, reflux time base, cooling temperature and heat rate.
(c) Samples were fractionated at two different reflux ratios (0.5 and 2) and at different reflux time bases (0.5 and 5 s). The fixed variables were pressure, charge, cooling temperature and heat rate.

3.5 Analysis and Tests Methods

The analysis and tests used for analyzing the oil samples to evaluate their properties were done according to the standard methods as shown in Table 3.3.

Table 3.3. Standard methods

Test	Designation	Apparatus
Relative density	ASTM D70 IP 190/86	Capillary-stoppered pyknometer
Viscosity		Schott Gerate (AVS 300)
Acid number	ASTM D974-86/IP 139/86	
Ash content	ASTM D482-80/IP 4/81	
Flash point	IP 35/86	Pensky–Martens open cup
Pour point	ASTM D97-78/IP 15/86	Denfuss
Saponification No.	ASTM D94-86/IP 136/86	
Viscosity index	ASTM D2270-87/IP 226/84	
Sulfur content		OXFORD (LAB-X 2000)
Metals content		Atomic absorption spectrophotometer (pye UNICAM sp9)

Chapter 4

The Re-refining Process Experimental Results

4.1 Dehydration

Several experiments were carried out for the dehydration of used lubricating oil. Those experiments were conducted by simple vacuum distillation. The results of these experiments are tabulated in Table 4.1, which shows that the best dehydration results are obtained at lower vacuum pressure and even though there is a wide range in boiling point between water, gasoline and the base oil cut. Also lower vacuum pressure is preferred to ensure that the temperature will not rise above 250 °C, which is the oil degradation temperature. The distribution of the water and gasoline in the used oil is not homogeneous which is indicated by the condensed liquid temperature. The final dehydration temperature depends on the amount of water and gasoline fractions in the used oil. The volume distilled is improved by decreasing the cooling temperature despite the fact that there is a wide range between the vapor temperature and cooling temperature. The amounts of water and gasoline separated in all runs were small due to low fuel dilution. The first experiment was considered the best dehydration condition.

4.2 Solvent Extraction

4.2.1 The Optimum Solvent to Oil Ratio

The solvent to oil ratio investigation is conducted at a solvent composition of 25% 2-propanol, 50% 1-butanol and 25% butanone as reported by Whisman et al. (1978). The results for mass balance for the optimum solvent to oil ratio experiments are tabulated in Table 4.2. While the tests for these experiments are tabulated in Table 4.3. The properties of produced solvent treated oil, i.e., oil recovery, ash reduction and sulfur reduction in relation to solvent to oil ratio are shown in Fig. 4.1.

The results of the investigation, Table 4.3 and Fig. 4.1 indicate that the maximum ash reduction is achieved for solvent to oil ratio of 4:1. The oil recovery and ash reduction for the same ratio are better than that obtained for solvent to oil ratio of 3:1 and 2:1. This indicates that by increasing the solvent amount, the solvency power is improved. The percentage of oil recovery for the solvent to oil ratio of 6:1 is further improved, but this solvent to oil ratio produces an ash reduction lower than that obtained for the solvent to oil ratio of 4:1 and 5:1 as shown in Fig. 4.1. That means that solvent to oil ratio larger than 4:1 will lead to dissolution of some contaminants in the solvent phase especially the ash forming material, which was considered to be undesirable. As a result of the above

Table 4.1. Dehydration data

Variable	Experiment						
	1	2	3	4	5	6	7
Pressure (mBar)	4	4	4	8	8	12	12
Final temperature (°C)	200	158	166	186	195	215	215
Volume distilled (mL)	25	15	22	10	15	5	10
Charge (liter)	1.5	1.5	1.5	1.5	1.5	1.5	1.5
First drop temperature (°C)	131	119	115	167	187	190	186
Cooling temperature (°C)	20	20	8	20	6	20	4

Table 4.2. Measurements of mass balance for optimum solvent to oil ratio experiments

Solvent to oil ratio	Oil feed (g)	Solvent (g)	Extract (g)	Raffinate (g)	Extract Oil (g)	Extract Solvent (g)	Loss (g)
2:1	45.11	79.29	108.95	15.45	39.35	69.50	0.10
3:1	45.10	117.26	150.14	12.22	40.18	109.70	0.27
4:1	45.00	158.12	192.80	10.32	42.77	149.85	0.18
5:1	45.00	197.53	237.50	5.04	42.72	194.61	0.17
6:1	45.00	239.20	276.90	7.30	43.27	233.82	0.00

Table 4.3. Test analysis of the optimum solvents to oil ratio experiments

Solvent to oil ratio	Oil recovery (wt%)	Solvent recovery (wt%)	Ash content (wt%)	Ash reduction (wt%)	Sulfur content (wt%)	Sulfur reduction (wt%)
2:1	87.22	87.65	0.272	22.73	1.389	14.73
3:1	89.10	93.55	0.207	41.93	1.422	12.71
4:1	95.05	94.77	0.163	53.69	1.483	8.96
5:1	94.94	98.52	0.194	44.89	1.238	24.00
6:1	96.15	97.75	0.279	20.80	1.213	25.54

mentioned facts, the solvent to oil ratio of 4:1 was considered to be the better solvent to oil ratio used for treatment of used lubricating oil. Nevertheless, ratios above 3:1 were not considered economically feasible by industry. Thus a solvent to oil ratio of 3 to 1 is considered the optimum ratio in this study because it gives good ash reduction, good oil recovery and low cost.

4.2.2 Optimum Solvent Composition

The results for optimum solvent composition were tabulated in Table 4.4, which represent the mass balance of the whole 15 experiments, while Table 4.5 represents the test data for the oil produced after solvent stripping. These data are the oil recovery, solvent recovery, ash content, ash reduction, sulfur content and sulfur reduction.

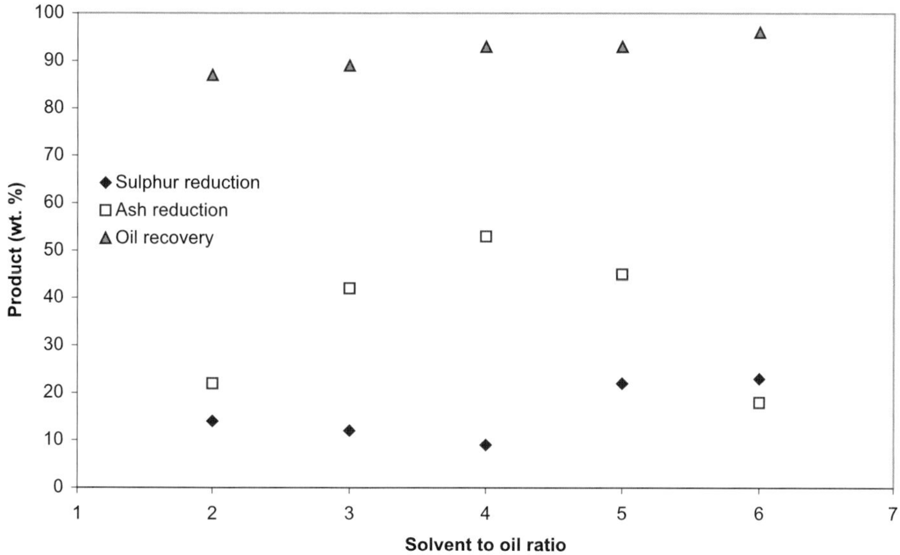

Fig. 4.1. Solvent to oil ratio vs. weight percent product.

Table 4.4. Measurements of mass balance for the optimum solvent composition experiments

Exp. No.	Oil feed (g)	Solvent (g)	Feed		Extract		Loss (g)
			Extract (g)	Raffinate (g)	Oil (g)	Solvent (g)	
1	36.5	94.00	122.90	7.60	34.30	88.37	0.21
2	36.02	92.02	115.10	13.30	28.68	86.55	0.00
3	36.02	91.90	120.10	8.00	34.45	85.78	0.00
4	36.02	93.10	119.50	9.80	29.98	89.50	0.02
5	36.00	92.80	123.40	5.40	34.73	88.50	0.18
6	36.20	95.30	126.50	5.00	32.67	93.82	0.00
7	36.80	95.00	125.30	6.50	35.06	90.19	0.05
8	36.20	91.90	120.50	7.60	32.71	87.53	0.26
9	36.80	95.17	123.17	8.80	33.68	89.47	0.02
10	36.88	94.18	115.98	15.00	28.25	87.36	0.37
11	36.80	96.11	123.17	9.80	33.26	89.86	0.05
12	36.20	93.80	124.80	5.20	31.31	93.31	0.18
13	36.02	92.60	118.90	9.90	31.31	87.56	0.03
14	36.40	94.30	122.6	8.10	32.36	90.30	0.00
15	36.30	93.6	122.80	7.10	31.32	91.50	0.00

4.2.2.1 *Model fitting.* The responses for the re-refining experiments were weight percentage oil recovery and weight percentage ash reduction. The oil recovery illustrates the economy of the process. While the ash reduction represents the amount of impurities removed by the process. Multivariable polynomial models were found to be best suited to all responses because they are simple and represent a reasonably accurate fit. In forming the

Table 4.5. Test analysis of the optimum solvent composition experiments

Exp. No.	Oil recovery (wt%)	Solvent recovery (wt%)	Ash content (wt%)	Ash reduction (wt%)	Sulfur content (wt%)	Sulfur reduction (wt%)
1	94.04	94.01	0.153	56.53	1.463	10.19
2	79.23	93.87	0.142	59.66	1.410	13.44
3	95.17	93.34	0.159	44.72	1.398	14.18
4	82.83	96.13	0.142	59.74	1.429	12.28
5	96.47	95.37	0.234	33.52	1.059	35.00
6	90.25	98.45	0.196	44.32	1.438	11.73
7	95.26	94.94	0.194	44.86	1.379	15.35
8	90.34	95.24	0.225	35.97	1.447	11.17
9	91.53	94.01	0.230	34.66	1.363	16.33
10	76.60	92.76	0.089	74.72	1.521	16.63
11	90.24	93.50	0.115	67.44	0.411	13.38
12	86.48	99.48	0.180	48.86	1.059	35.00
13	86.50	94.56	0.250	28.98	1.330	18.36
14	88.89	95.76	0.261	25.85	1.449	11.05
15	86.51	97.76	0.206	41.31	1.439	11.66

polynomial equation, it is found that by replacing the response of y by $\ln y$ accuracy was improved. The response surface for oil recovery ($\ln y_1$) and ash reduction ($\ln y_2$) were fitted by the least squares method to estimate polynomial regression coefficients. The developed two equations are reported below:

(1) For oil recovery

$$\begin{aligned}\ln y_1 = {} & 48.61 + 0.378x_1 + 1.676x_2 + 3.215x_3 - 8.376 \times 10^{-3}x_1^2 - 2.109 \times 10^{-2}x_2^2 \\ & - 3.644 \times 10^{-2}x_3^2 - 2.936 \times 10^{-2}x_1x_2 - 4.455 \times 10^{-2}x_1x_3 \\ & - 5.788 \times 10^{-2}x_2x_3. \end{aligned} \tag{4.1}$$

(2) For ash reduction

$$\begin{aligned}\ln y_2 = {} & -75.169 - 0.747x_1 + 0.641x_2 + 1.263 \times 10^{-2}x_1^2 - 2.362 \times 10^{-2}x_3^2 \\ & + 1.857 \times 10^{-2}x_1x_2 + 4.089 \times 10^{-2}x_1x_3 + 2.080 \times 10^{-2}x_2x_3 \\ & - 3.061 \times 10^{-4}x_1x_2x_3 + 4.689 \times 10^{-4}x_3^3. \end{aligned} \tag{4.2}$$

Two terms of third order polynomials were added in Eq. (4.2) in order to minimize both the average absolute error and standard deviation. Figures 4.2 and 4.3 show the relationship between calculated and experimental results for each response. While Table 4.6 represents the statistical analysis of Eqs. (4.1), (4.2).

4.2.2.2 *The optimization procedure.* The optimum solvents composition is the solvents composition that provides highest oil recovery and ash reduction. The operating ranges

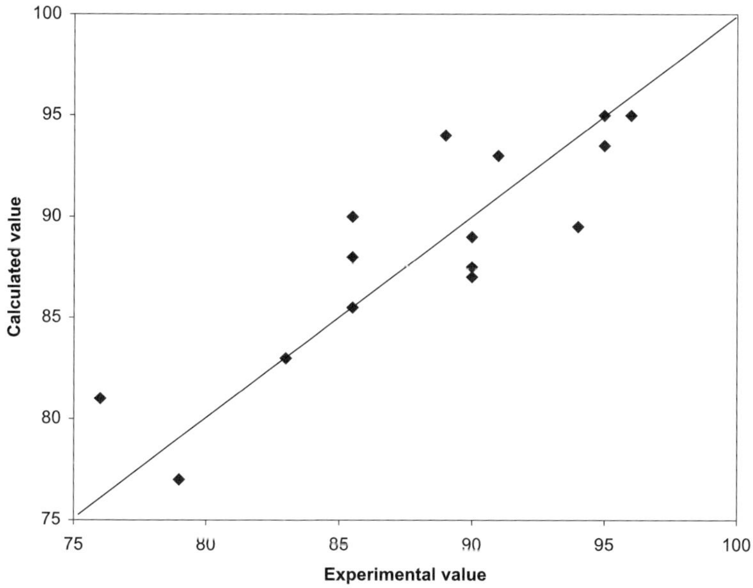

Fig. 4.2 Experimental values vs. calculated values for Eq. (4.1).

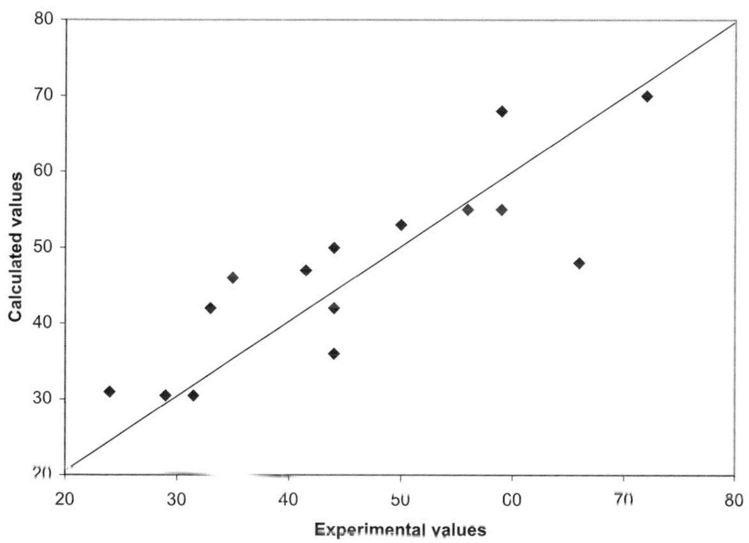

Fig. 4.3. Experimental values vs. calculated values for Eq. (4.2).

for the experiment variables represent constraints to the functions of oil recovery and ash reduction. These inequality constraints are:

$$10\% \leqslant x_1 \leqslant 40\%, \tag{4.3}$$

Table 4.6. Statistical analysis of Eqs. (4.1) and (4.2)

Function	Eq. (4.1)	Eq. (4.2)
Average absolute error	2.466%	14.476
Correlation coefficient	0.999	0.966
F test value	286.429	7.732
Estimated standard deviation	4.549	14.431

$$35\% \leqslant x_2 \leqslant 65\%, \tag{4.4}$$

$$10\% \leqslant x_3 \leqslant 40\%. \tag{4.5}$$

The summation of the three variables, which must be 100%, is an equality constraint:

$$x_1 + x_2 + x_3 = 100\%. \tag{4.6}$$

The Interior Penalty Function Method (Rao, 1979) can find the optimum point for a function subjected to mixed equality and inequality constraints. To simplify the solution, a maximization problem can be turned to a minimization problem by changing the function sign. The optimization problem is formulated as minimize $F(X) = -(\ln y_1 + \ln y_2)$ subjected to

$$G_j(X) \leqslant 0, \quad j = 1, 2, 3, \ldots, m,$$

$$L_j(X) = 0, \quad j = 1, 2, 3, \ldots, p,$$

where X is the vector of process variables, G_j is set of m inequality constraints, L_j is set of p equality constraints.

This constrained minimization problem can be converted to an unconstrained minimization problem by constructing a function of the form

$$\phi_r = \phi(X, R_r) = F(X) - R_r \sum_{j=1}^{m} \frac{1}{G_j(X)} + \frac{1}{\sqrt{R_r}} \sum_{j=1}^{p} L_j^2. \tag{4.7}$$

This equation was reported by Fiacco and McCormic (Rao, 1979), R_r is a positive constant known as the penalty parameter. The unconstrained minima $\phi(X, R_r)$ defined by Eq. (4.7), for a sequence of values $R_1 > R_2 > \cdots > R_r$ converges to the optimal solution of the constrained problem as $R_r \to 0$.

The ϕ_r for this case is

$$\begin{aligned}\phi_r = \phi(x_1, x_2, x_3, R_r) &= F(x_1, x_2, x_3)\\ &- R_r\left[\frac{1}{x_1 - 10} + \frac{1}{40 - x_1} + \frac{1}{x_2 - 35} + \frac{1}{65 - x_2} + \frac{1}{x_3 - 10} + \frac{1}{40 - x_3}\right]\\ &+ \frac{1}{\sqrt{R_r}}[x_1 + x_2 + x_3 - 100]^2.\end{aligned} \tag{4.8}$$

The initial base point X_1 is selected to satisfy the inequality constraint and not necessarily satisfy the equality constraint. The unconstrained optimum point was evaluated by Hook and Jeeves' method (Bunday, 1984). The algorithm as shown in Fig. 4.4 explains this optimization technique as a whole. A program was developed in order to undertake the optimization procedure. The program was written in the Turbo-Basic programming language and a copy of the program is shown in Appendix B. The value of R_1 was chosen to be 45 and the accuracy (final value of R_r) set to be 1×10^{-8}. The optimization procedure is described as follows (Awaja and Pavel, 2005):

(1) The value of R_r is 45 for this study and is set as an input to the program. The next value of R_r is calculated according to the equation $R_{r+1} = C \times R_r$, where C is a constant and its value is less than 1; a value of $C = 0.1$ is used for this study.
(2) The step length (H) also has been set as an input; a value between 0 and 1 is used and it depends on the sensitivity of the minimum point. Different step lengths could be selected for each variable, however in this optimization problem one initial step length is selected for both variables, with a value of 0.1.
(3) A base point (X_1) is selected based on medium values for the two variables and in satisfaction of the constraints.
(4) Exploration about the base point (X_1) is initiated. That means that the function value at the base point and its neighborhood is explored. The exploration of the function neighborhood provides knowledge that helps to make a pattern move in the next stage that results in a greater reduction in the function value. The exploration process takes place as follows:
   (a) The function value $P(R_1, X_1) = P(R_1, x_{11}, x_{21}, x_{31})$ is calculated.
   (b) Each variable, in turn, is modified by the step length.
   (c) The function value $P(R_1, x_{11} + H, x_{21}, x_{31})$ is evaluated. If

$$P(R_1, x_{11} + H, x_{21}, x_{31}) < P(R_1, x_{11}, x_{21}, x_{31}),$$

   then the base point is replaced with a new one $X_2 = (x_{11} + H, x_{21}, x_{31})$; otherwise,
   (d) The function value $P(R_1, x_{11} - H, x_{21}, x_{31})$ is evaluated. If

$$P(R_1, x_{11} - H, x_{21}, x_{31}) < P(R_1, x_{11}, x_{21}, x_{31}),$$

   the base point is replaced by $X_2 = (x_{11} - H, x_{21}, x_{31})$; otherwise,
   (e) The modification in x_2 and x_3 variable is considered and step (c) is repeated.
   (f) After a function reduction involving exploration of the three variables, the base point is changed to new one (X_2), which is generated from adding or subtracting a step length.
   (g) If no function reduction is achieved after modifying both variables, i.e. $X_1 = X_2$, the step length is reduced to one tenth of its present value and the exploration procedure is repeated.
   (h) If a function reduction is achieved, i.e. X_2 not equal to X_1, a pattern move is then made.

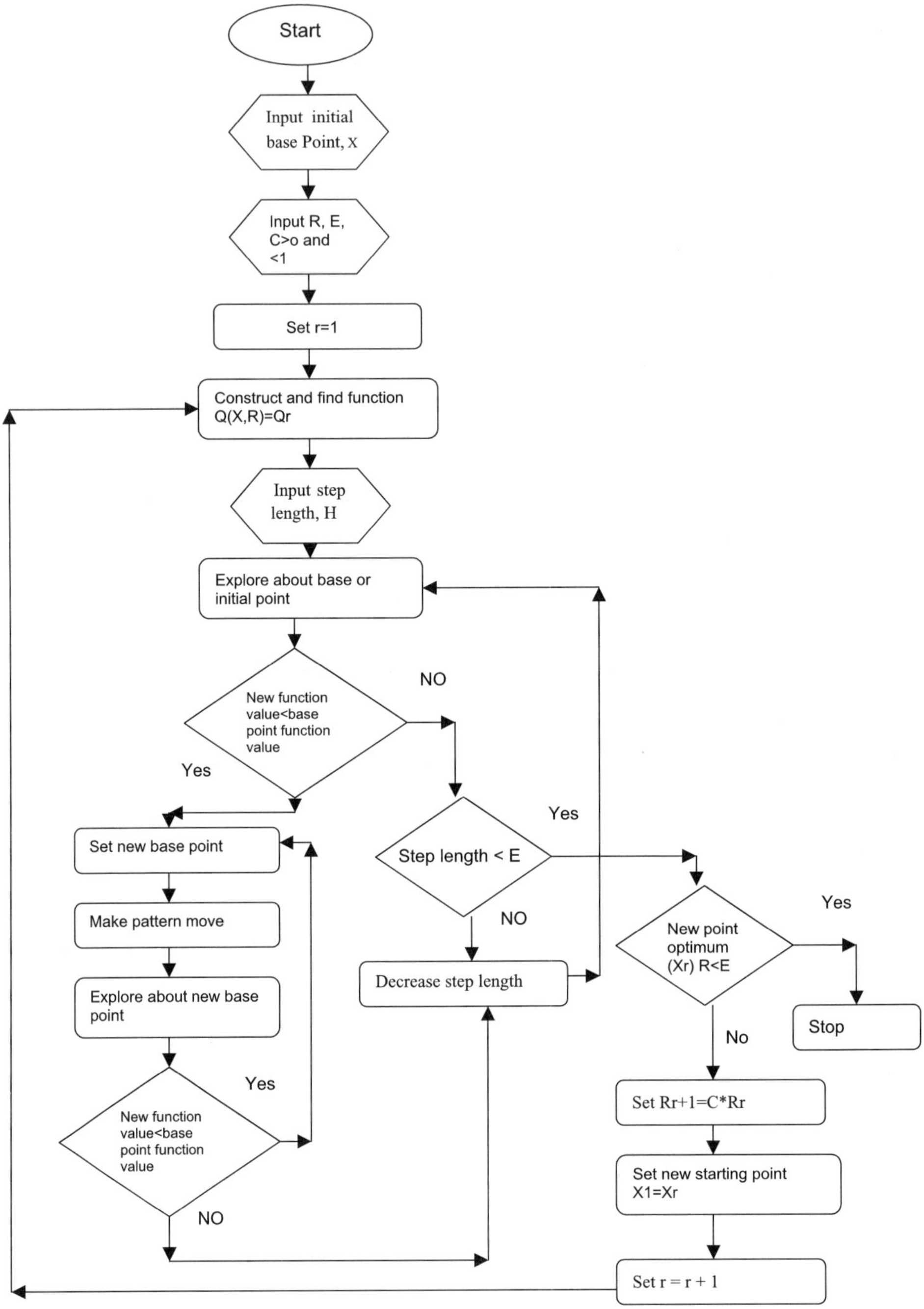

Fig. 4.4. The optimization algorithm.

(5) The knowledge gained from the exploration step is used to make the search move towards the function minimum direction (pattern) instead of searching in all directions. The pattern move procedure takes place as follows:
   (a) A new base point (X_3) is generated, where

$$X_3 = X_1 + 2(X_2 - X_1),$$

$$(x_{13}, x_{23}, x_{33}) = (x_{11}, x_{21}, x_{31}) + 2((x_{13} - x_{12} - x_{11}), (x_{23} - x_{22} - x_{21})).$$

   (b) Exploration of X_3 is evaluated.
   (c) If a further reduction in function value is achieved, then step (a) is repeated using the new base point $X_4 = X_2 + 2(X_3 - X_2)$, otherwise the pattern move is discontinued and exploration about the older base point takes place.

(6) The exploration and pattern move processes are terminated when a step length is reduced to an accuracy value (E). The accuracy value for this study is set to a value of 1×10^{-8}.

After implementing the function optimization, the following results (Table 4.7) were obtained for different selected initial points.

The first optimum point was chosen to be the optimum point for the process and its solvent composition was tested experimentally and gives an ash reduction of 65% and oil recovery of 96%.

The capability of a polar solvent to segregate sludge (contaminants) from used oil is closely related to its solubility parameters. The flocculation action of a polar solvent is correlated to the difference between solubility parameters of the solvent (S_1) and polyisobutylene (S_2) used as viscosity improver in the lubricating oil additives package (Reis and Jeronimo, 1988, 1990). The most efficient solvent mixture used for purification of used oil must be the highest solubility parameter. 2-propanol has a solubility parameter difference with polyisobutylene of about (7.4) that is larger than 1-butanol of (7.0) and both alcohols are much larger solubility difference than butanone of (2.8).

It was recommended by Whisman et al. (1978) to use solvent treatment mixture containing 25% 2-propanol, 50% 1-butanol and 25% butanone. They also reported that using higher percentage of 2-propanol than 25% produces poorly separated sludge. While using higher percentage of butanone than 25% produces good sludge separation, but it dissolves ash-forming compounds. The solvent mixture used by Whisman et al. (1978) shows that 1-butanol was used as a base solvent as indicated by its higher percentage in the solvent mixture. They chose an equal percentage of 2-propanol and butanone to moderate the ash reduction and oil recovery properties of 1-butanol. The solvent mixture used by Whisman

Table 4.7.

Optimum point	x_1 (%)	x_2 (%)	x_3 (%)	Ash reduction (%)	Oil recovery (%)
1	25	35	40	68.95	98.83
2	40	35	25	91.85	79.51

et al. (1978) is very selective since it contains 75% of high solubility parameter solvents. The reported ash reduction of 75% indicates that the original used oil treated was highly contaminated. The 25% of butanone was used to improve the oil recovery amount since butanone has a lower solubility parameters value than the two alcohols.

An investigation of the suitability of the Whisman et al. (1978) solvent composition was carried out. It was found that 25% 2-propanol, 35% 1-butanol and 40% butanone was the most suitable solvent composition to be used with Iraqi used lubricating oil. This solvent mixture produced 96% oil recovery and 65% ash reduction.

It was found that some oil components, which have high molecular weights, in addition to the contaminants that have also high molecular weights, do not dissolve in rich propanol solvent mixtures, since 2-propanol is very selective to the high molecular weight oil components. Consequently valuable high molecular weight oil component will not be recovered and lost to the raffinate phase. Thus, these components must be re-extracted from the contaminants.

Rich 1-butanol and butanone solvent mixture will dissolve base oil components containing some contaminants. These contaminants must be removed from the extract, therefore these mixtures are shown to be inefficient. Nevertheless, the experiments indicate that 1-butanol and butanone act together as a power solvent, as indicated by the oil recovery values.

Tables 4.4 and 4.5 indicate that increasing the percentage of 2-propanol above 25% produce higher ash reduction, higher sludge removal and poor oil recovery, while increasing the percentage of butanone above 25% produces lower ash reduction, lower sludge removal and higher oil recovery. Therefore an efficient solvent mixture was expected with both 2-propanol and butanone has an amount above 25%.

1-butanol interacts with both 2-propanol selective action and butanone power action. The ash reduction and oil recovery depend on the percentage of 1-butanol in the solvent mixture. Lower percentages of 1-butanol than 50% with higher percentages of 2-propanol than 25% improve ash reduction. While higher percentages of 1-butanol than 50% with higher percentages of butanone than 25% improve oil recovery. In the extraction process, the vital parameter is ash reduction in order to produce oil with high purity without further separation processes; in the distillation process, oil recovery is the affecting parameter that insures minimum losses in the oil. In present research where solvent extraction is followed by vacuum distillation, optimization between oil recovery and ash reduction is required.

Optimum results showed that the optimum solvent mixture that produces maximum oil recovery (96%) with maximum ash reduction of 65%, is 25% 2-propanol, 35% 1-butanol, 40% butanone. Maximum ash reduction was produced (i.e., 92%) when the solvent composition is 40% propanol, 35% 1-butanol and 25% butanone. It is obvious that the optimum solvent composition differs from that composition reported by Whisman et al. (1978) mentioned above. This difference might be attributed to the nature of the base oil crude; which is a mixed base for Iraqi crude, and to the different components used in the additive package where in nowadays ash-less materials have been used instead of ash forming materials. The optimum solvent composition mixture contains 60% of high solubility parameter solvent.

This percentage produces an ash reduction of 65%. The 40% of butanone, which represent the maximum of butanone studied, adjusts the oil recovery value to 96%. Higher percentage of alcohols than 60% will improve the ash reduction especially when the percentage of 2-propanol is increased. However, the oil recovery will decrease to the extent that it will be unfeasible. The maximum ash reduction produced was formed by mixture of 40% 2-propanol, 35% 1-butanol and 25% butanone. The presence of 75% alcohols in the mixture with higher amount of 2-propanol than 1-butanol makes the mixture very selective and produces ash reduction of about 92% and oil recovery of 80%. This oil recovery was considered unfeasible.

Further inspection of Tables 4.4 and 4.5 indicates that increasing the percentage of 2-propanol above 25% produce higher ash reduction, higher sludge removal and poor oil recovery, while increasing the percentage of butanone above 25% produces lower ash reduction, lower sludge removal and higher oil recovery. Therefore an efficient solvent mixture was expected when both 2-propanol and butanone has a percentage higher than 25%.

After the optimum solvent to oil ratio of 3 to 1 was obtained and optimum composite solvent was established, which is 25% by volume 2-propanol, 35% by volume 1-butanol and 40% by volume butanone. The solvent treatment according to these optimum values of large quantities of dehydrated used oils was conducted.

The amount of dehydrated used oil charge was around 7.75 kg and the amount of solvents was 20.35 kg. These were mixed and separated to produce an extract amount of 26.65 kg and raffinate amount of 1.45 kg for 34 runs.

4.3 Solvent Stripping

Two different processes were used for solvent stripping, namely simple vacuum distillation as shown in Fig. 3.4 and simple atmospheric distillation as shown in Fig. 3.5. The most serious problem that appeared in solvent stripping was the foaming of the oil–solvent mixture. The foaming problem led to the carry over of the liquid mixture from the still pot to the column, to the condenser, and at the same time to the vacuum lines. The foaming problem was affected by the following:

(1) Heat rate: In order to define optimum heat rate, different heat rates were investigated, i.e. 330 W, 340 W, 380 W, 410 W, 435 W, 460 W and 490 W. It was found that the best heating rate with no foaming was around 380–410 W. Higher heat rate than 410 W caused severe foaming, therefore it is important to hold the heat rate constant. Heating rate is an important factor to prevent foaming. Since a high sudden heating rate accelerates foaming, while slow gradual heating rate reduces foaming.
(2) Pressure: The minimum pressure that was used without causing foaming is 120 mBar. Lower pressure than 120 mBar resulted in suction of the liquid to the vacuum lines.
(3) Composition: It was noticed that 2-propanol and butanone solvents are the cause of foaming, because when they are removed no foams appeared.
(4) Quantity: The charge above 1.5 litres (3/4 of the volume of still pot) will cause foams because of the reduction in space above the liquid surface that does not allow foam to subside.

Vacuum distillation was conducted under two different vacuum pressures of 120 mBar and 160 mBar. The results are shown in Figs. 4.5–4.8. Vacuum pressure is influential to

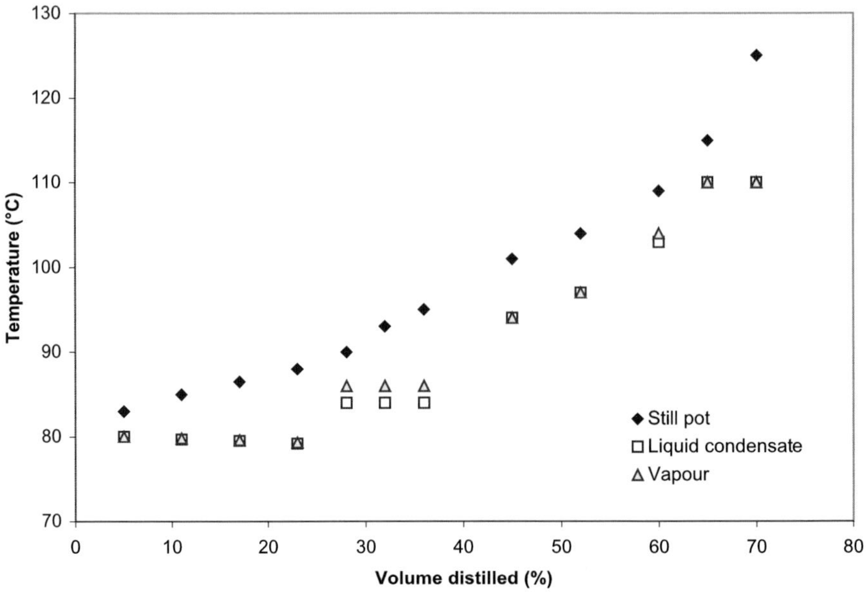

Fig. 4.5. Volume distilled vs. temperature for solvent stripping operating at 160 mBar. Charge is 1750 mL.

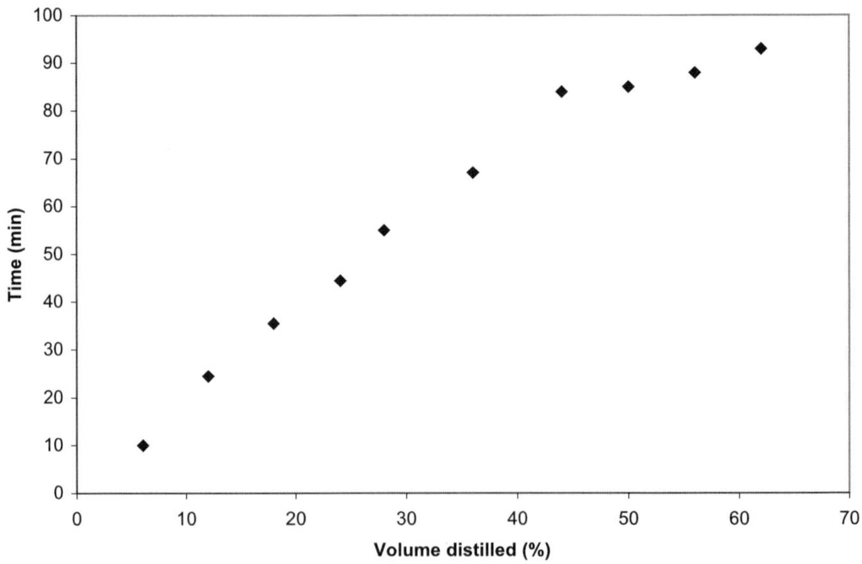

Fig. 4.6. Volume distilled vs. time of solvent stripping at 160 mBar pressure. Charge is 1750 mL.

the foaming phenomenon. The pressure in the vacuum distillation apparatus increases to more than that specified value, such as 120 mBar due to evaporation caused by higher

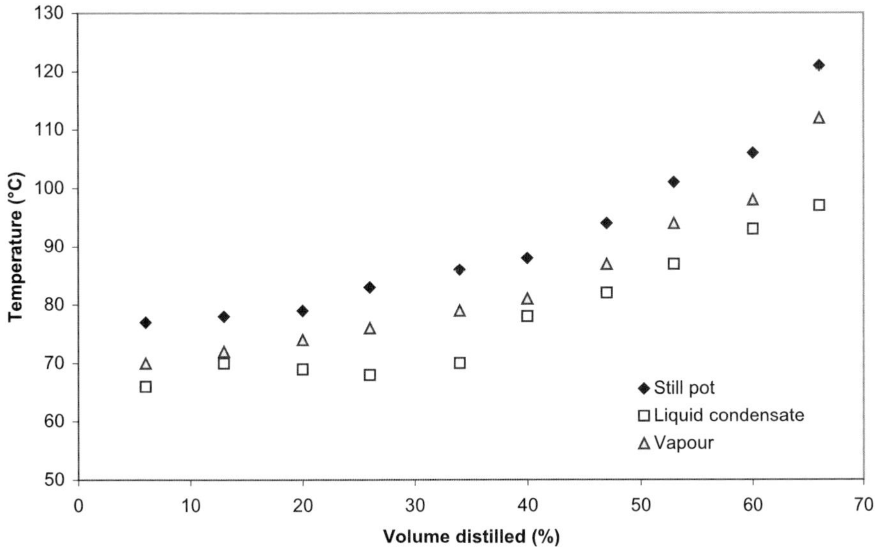

Fig. 4.7. Volume distilled vs. temperature of solvent stripping at 120 mBar and 1500 mL charge.

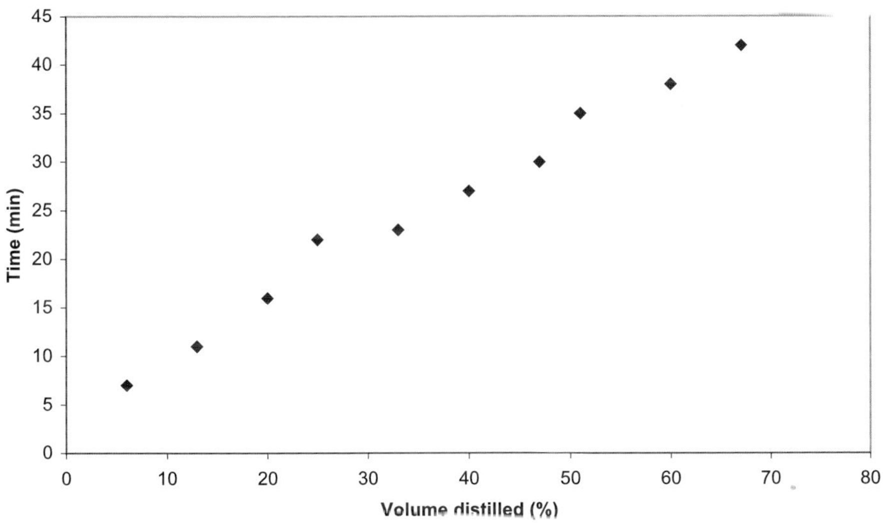

Fig. 4.8. Volume distilled vs. time of solvent stripping at 120 mBar and 1500 mL charge.

heating rate. The controller valve opens to stabilize the pressure by rejecting vapors out of the distillation unit. This will cause a turbulence that leads to an increase in foaming. Foaming was generally accompanied by severe carry over of liquid. The foaming is complicated by vacuum distillation due to the addition of another variable such as sucking or evacuation of vapors by the vacuum pump. To overcome this problem, a lower heating rate must be used. Figures 4.5–4.8 show that initially, temperature rise increases slightly due to removal of butanone alone from the mixture with no noticeable effect in the thermodynamic properties of the mixture. After that, it seems that there are thermodynamic properties of the mixture that differ from initial stage due to butanone removal from the mixture. These thermodynamic properties differences from initial stage are due to both butanone removal from the mixture and the structural configurations of butanone, 2-propanol and 1-butanol in the mixture, which shift the boiling point of the mixture to almost double its value after removing butanone. Higher increase in temperature was required in order to evaporate the 2-propanol and 1-butanol solvent from oil–solvent mixture, as shown in Fig. 4.5. The remaining 25% by volume is oil, which can vaporize in the above-mentioned temperature.

Figure 4.6 shows the time required to obtain different percentage volume of distillate. The rate of distillation initially was high and then decreases due to interference of oil molecules impregnated solvent. Lower vacuum pressure and lower feed charge gave better behavior for still pot temperature, liquid condensate temperature and vapor temperature with percentage of volume distilled, as shown in Fig. 4.7 compared with temperatures reported in Fig. 4.5. Lower vacuum pressure alone improves temperature and time, while lower feed charge improves time only. Similarly, the two variables of charge and vacuum pressure gave similar behavior by considering each variable alone. Consequently, both of the above variables were considered together with temperature as shown in Fig. 4.7 and with time as shown in Fig. 4.8. However, foaming problems arise without carry over at low vacuum pressure, i.e. 120 mBar.

The solvent stripping was conducted also by atmospheric distillation. Simple distillation gave satisfactory results, while distillation using a sieve plate column as shown in Fig. 3.6 was unpractical because the stripping takes a longer time with no solvent purity improvement. Since there are three solvents in the oil with large differences in boiling point temperature of two of the solvents, i.e., butanone (80 °C) and 2-propanol (82 °C), compared with the third solvent 1-butanol (117 °C). Heating to about 90 °C at atmospheric pressure will cause the more volatile component of the low boiling point of those solvents to vaporize, carrying with them the liquid of the third solvent. This creates a big foaming problem during operation at heat rate of 600 W per 1.5 liter. The foaming problem can be controlled by reducing the heat rate to 380 W. All atmospheric distillation experiments failed to recover all the solvent amounts at temperature below 250 °C, which is the degradation temperature of the oil, while butanone and 2-propanol were successfully recovered at 200 °C.

The optimum procedure is a combination of atmospheric distillation and vacuum distillation. Initially atmospheric distillation was used at a heat rate of 380 W to recover butanone and 2-propanol solvents followed by vacuum distillation down to 40 mBar to recover the 1-butanol solvent. No foaming problems were noticed during the recovery of the 1-butanol solvent. This operation shows that no anti-foaming material was needed during stripping.

According to the above optimum procedure, 25.75 kg of solvent–oil mixture (extract) was stripped to produce 7.5 kg of solvent treated oil and 18.1 kg of recovered solvent with 150 g of lost solvents.

4.4 Vacuum Distillation

Fractionation behaviors of lubricating oils under vacuum distillation were studied by plotting the still pot temperature, liquid condensate temperature, vapor temperature, time, total vapor velocity and fraction vapor velocity against the percentage of volume distilled. While in some cases the total vapor velocity and fractions vapor velocity were plotted against still pot temperature.

The overall general behaviors of used, virgin and solvent treated oil fractionations are believed to be as follows. The still pot temperature increases rapidly initially with volume distilled due to the presence of different light components having different boiling points. Then the still pot temperature increases gradually due to the presence of different fractions of components having small variations in boiling point. Finally, the still pot temperature increases rapidly at the end due to the presence of heavy components having a wide range of boiling points. Since the light components were removed by a dehydration step in the initial stage of processing, fast initial stage of solvent treated oil distillation was reduced. The above-discussed general behaviors are described in Figs. 4.9, 4.15, 4.21, 4.27, 4.33, 4.39, 4.48, 4.54, 4.60 and 4.66.

Generally liquid condensate temperature and vapor temperature have different behaviors to still pot temperature. They initially increase with the increase of percentage volume distilled. This might be explained as being due to higher vapor load compared to liquid condensed. Then, after they reach maximum value the liquid condensate temperature and

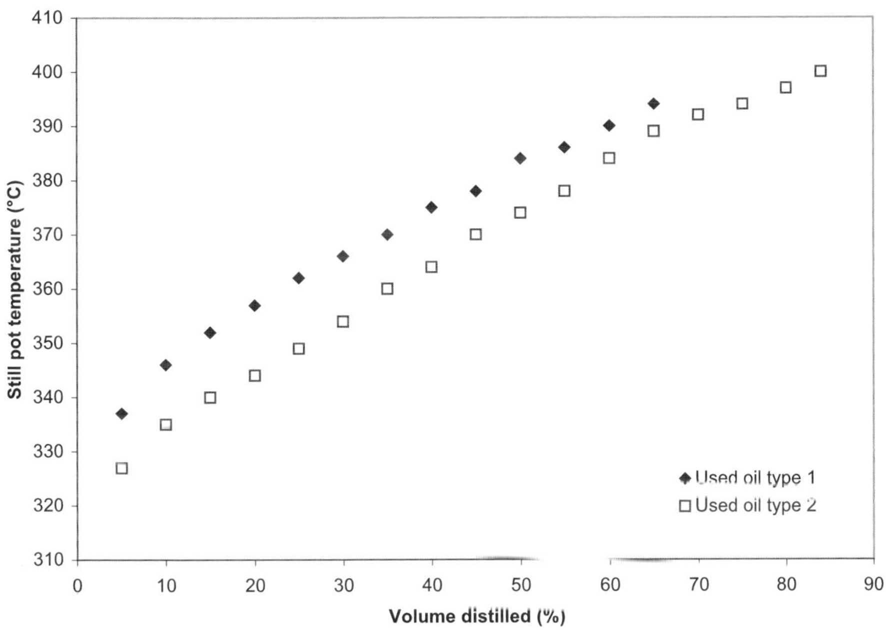

Fig. 4.9. Volume distilled vs. still pot temperature for different types of used oils. Reflux time base is 0.5 s, $R_r = 2$, heat rate = 600 W, cooling temperature = 15 °C, charge = 1000 mL, and $P = 8$ mBar.

vapor temperature decrease with increasing percentage volume distilled as a result of low vapor load compared with liquid condensation load as shown in Figs. 4.10, 4.11, 4.16, 4.17, 4.22, 4.23, 4.28, 4.29, 4.34, and 4.35. The behavior of liquid condensate temperature and vapor temperature has some changes for fractionation of solvent treated oil. The region

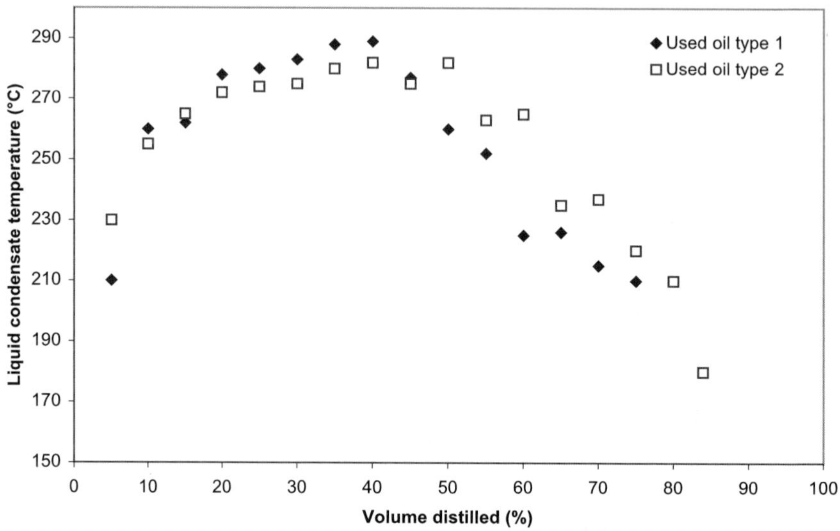

Fig. 4.10. Volume distilled vs. liquid condensate temperature for different types of used oils. Reflux time base is 0.5 s, $R_r = 2$, heat rate = 600 W, cooling temperature = 15 °C, charge = 1000 mL, and $P = 8$ mBar.

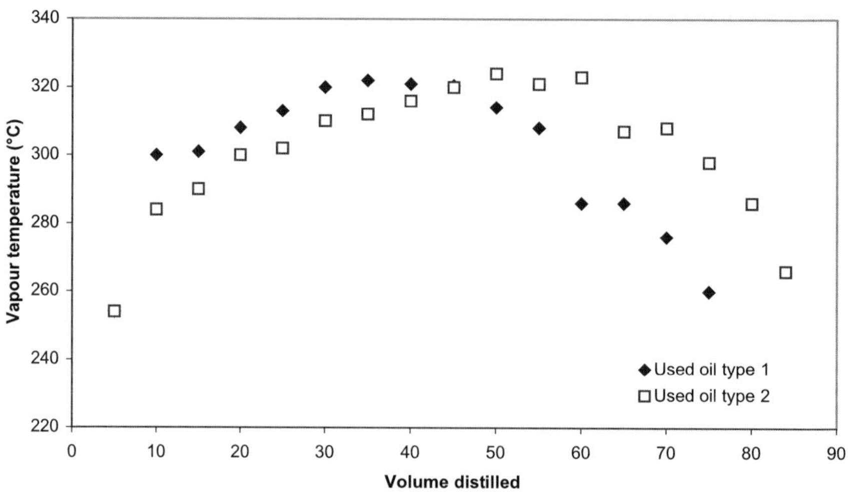

Fig. 4.11. Volume distilled vs. vapor temperature for different types of used oils. Reflux time base is 0.5 s, $R_r = 2$, heat rate = 600 W, cooling temperature = 15 °C, charge = 1000 mL, and $P = 8$ mBar.

of negative gradient of the curves becomes smaller as shown in Figs. 4.40, 4.41, and 4.49. The results for each type of lubricating oils fractionation are as follows.

4.4.1 Used Lubricating Oil

Experiments on used lubricating oil fractionations were carried out by studying the following.

4.4.1.1 *Two types of used lubricating oils were fractionated i.e. type 1 is used oil drawn from one automobile while type 2 is used oil mixture drawn from many automobiles.* The experiments were conducted at 8 mBar vacuum pressure and reflux ratio of 2. Figures 4.9–4.14 and Tables 4.8 and 4.9 represent the two types distillation behaviors and their distillates properties.

Table 4.8 shows that used oil type 2 produced higher lubricating oil base than used oil type 1. The two used oil types have different properties, i.e. used oil type 1 has higher amount of heavy components than type 2 as shown in Table 4.9 indicated by the higher viscosity, higher flash point, and lower pour point. The distillates of used oil type 2 produce better viscosity and flash point than distillates of used oil 1, but both distillates are not suitable for reuse, since they produce high ash and sulfur content. Even though they have acceptable viscosity index and pour point.

The distillation behaviors for the two used oil types, Figs. 4.9–4.14 indicate that used oil type 2 produced lower distillation temperatures and a higher distillation rate than type 1. The variation of still pot temperature with percentage volume distilled depends on the type and component of the used lubricating oil. The percentage volume distilled increased with still pot temperature, but the rate of increase in volume distilled decreased with still pot temperature, compared with the initial rate as shown in Fig. 4.9.

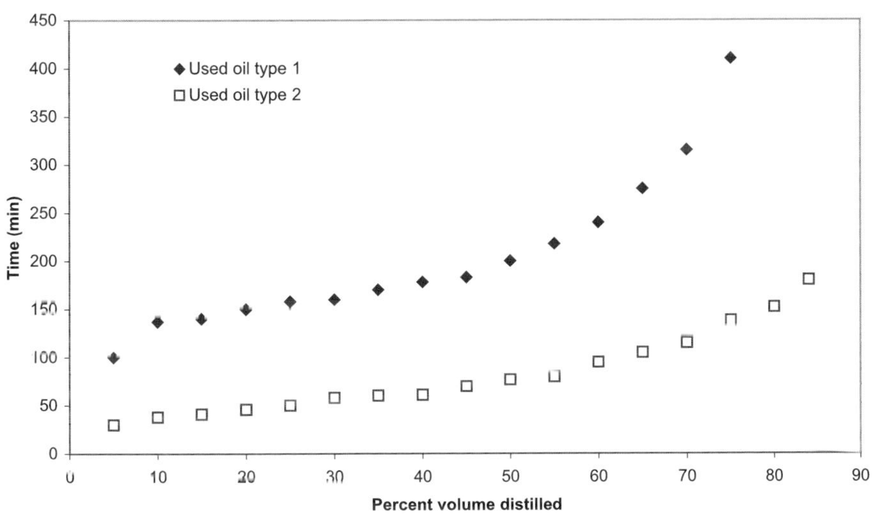

Fig. 4.12. Volume distilled vs. time for different types of used oils. Reflux time base is 0.5 s, $R_r = 2$, heat rate = 600 W, cooling temperature = 15 °C, charge = 1000 mL, and $P = 8$ mBar.

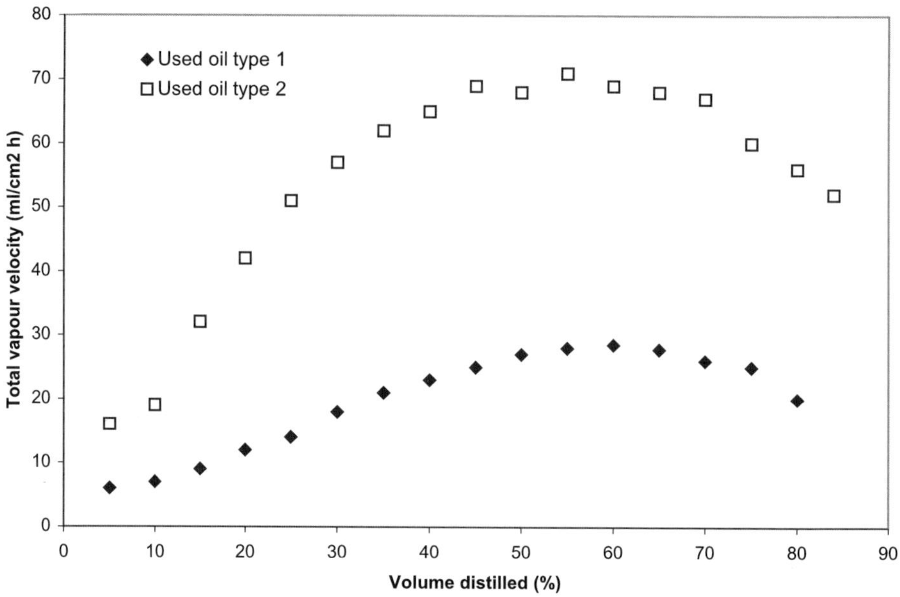

Fig. 4.13. Volume distilled vs. total vapor velocity for different types of used oils. Reflux time base is 0.5 s, $R_r = 2$, heat rate = 600 W, cooling temperature = 15 °C, charge = 1000 mL, and $P = 8$ mBar.

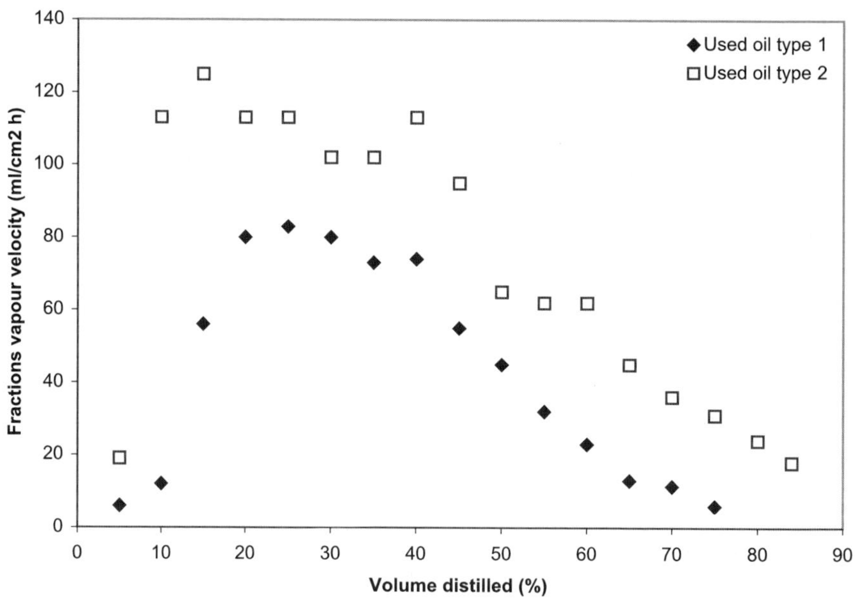

Fig. 4.14. Volume distilled vs. fractions vapor velocity for different types of used oils. Reflux time base is 0.5 s, $R_r = 2$, heat rate = 600 W, cooling temperature = 15 °C, charge = 1000 mL, and $P = 8$ mBar.

Table 4.8. Fractions obtained for distillation of the two types of used oil at 8 mBar pressure, reflux ratio of 2, and 400 °C final temperature

Fraction	Used oil type (1) (vol%)	Used oil type (2) (vol%)
Gasoline and water	3.4	4
Oil base, first cut	71.2	50
Oil base, second cut	–	29
Residue	25.4	17

Table 4.9. Tests analysis for the two used oils and their distillates at 8 mBar pressure, reflux ratio of 2 and 400 °C final temperature

Test	Used oil type 1	Used oil type 2	Used oil type 1 distillates	Used oil type 2 distillates	
			Oil base 1	Oil base 1	Oil base 2
Sp gr	0.900	0.905	0.899	0.903	0.914
Viscosity 40 °C (cSt)	84.3	71.1	20.8	39.3	36.1
Viscosity 100 °C (cSt)	10.0	9.0	4.6	6.4	5.7
Flash point (°C)	183	155	152	195	180
Pour point (°C)	−17	−6	−9	−5	1
Ash (wt.%)		0.35	0.11	0.26	0.13
Carbon (wt.%)		2.25			
Sulfur (wt.%)		1.67	1.35	1.14	1.33
Acid No.		2.55			
Saponification No.		11.2			
Water (wt.%)		2			
Benzene insoluble (wt.%)		0.02			
Refractive index			1.497		
Viscosity index	98	100	144	112	95

Used oil type 2 is a better alternative than used oil type 1, since it can be distilled at still pot temperature lower than used oil type 1. Figure 4.10 illustrates the liquid condensate temperature with percentage volume distilled. The liquid condensate temperature increased with volume percentage distilled due to higher vapor load until reaching a maximum at 25% to 40% volume distilled. After that the liquid condensate temperature decreases with percentage volume distilled due to loss of sensible heat of the heavy ends to the cooling system. The two used oil type's curves coincide with one another because the used oil type 2 has higher vapor load with low vapor temperature, while used oil type 1 has low vapor load with higher vapor temperature. The pattern that initiated by an increase in vapor temperature with percentage volume distilled to maximum and then decrease as shown in Fig. 4.11 is explained as due to different types of oil component of used lubricating oil. The decrease of vapor temperature with percentage volume distilled is due to vapor load. Figure 4.12 indicates that used oil type 2 contains lighter oil components than with used oil type 1, which contain heavy oil components as indicated by the fact that used oil type 2 required a shorter time to be distilled compared with used oil type 1. Vapor velocity of used oil type 2 is more than that of used oil type 1 as shown in Figs. 4.13 and 4.14.

For the reasons discussed above, it was found out by experiments that used oil type 2 is better than used oil type 1, therefore further investigation continued with used oil type 2 only.

4.4.1.2 *Vacuum pressure.* Fractionation of used oils was carried out at different pressures namely 4 mBar and 8 mBar. Figures 4.15–4.20 and Tables 4.10 and 4.11 report the effects of the two vacuum pressures on both fractionation behavior of used oil and properties of the distillates.

Tables 4.8 and 4.10 show that the lubricating oil base amount obtained at 4 mBar pressure was lower than 8 mBar pressure. Tables 4.9 and 4.11 indicate that the distillates of used oil under 4 mBar pressure produces better ash content, Sp gr, and pour point than that of 8 mBar pressure. The oil base cut for 4 mBar pressure has better oil properties than that of 8 mBar pressure, but both of them were not suitable for re-use. The distillation behaviors are shown in Figs. 4.15–4.20, which indicate that 8 mBar pressure produces a better distillation rate than 4 mBar pressure since 4 mBar pressure cause sucking of the vapor up to 18.5%. Figures 4.16–4.18 indicate that higher distillation results were obtained using 4 mBar pressure compared with 8 mBar pressure. At 4 mBar pressure 18.5% of the vapor was sucked by the vacuum unit and the vapor did not reach the condenser by sucking it from the reflux tubes, therefore these above figures can be ignored. Still pot temperature under 8 mBar pressure gives high distillation results than 4 mBar pressure as shown in Fig. 4.15. This was not expected to give better results than lower vacuum pressure. This can be explained due to the vapor sucking at 4 mBar pressure. That leads to shift the entire curve by about 18.5%.

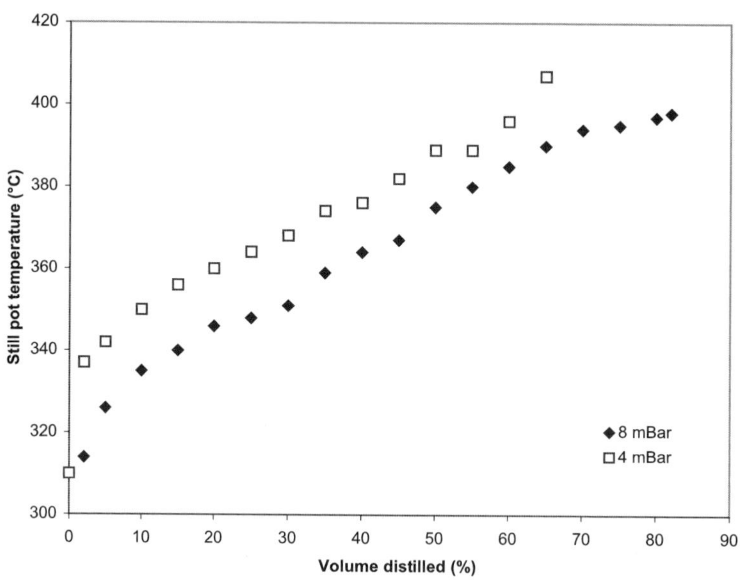

Fig. 4.15. Volume distilled vs. still pot temperature of used oil at different pressure. Reflux time base is 0.5 s, $R_r = 2$, heat rate = 600 W, cooling temperature = 15 °C.

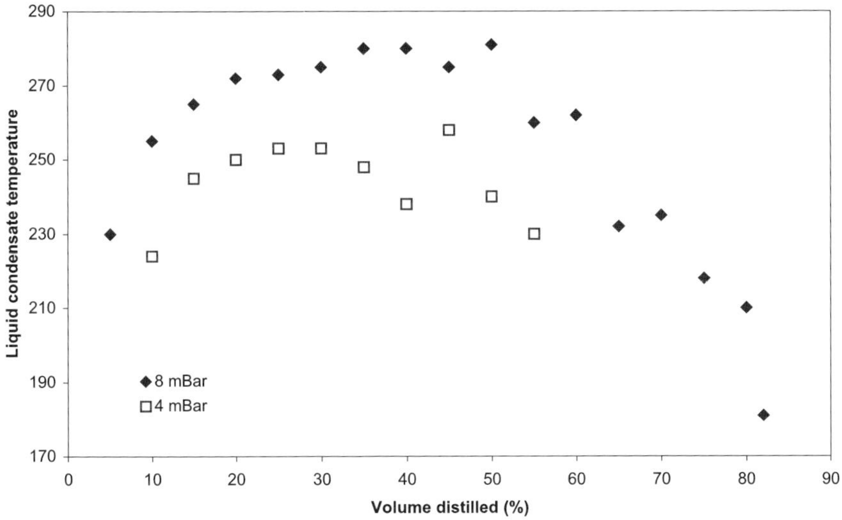

Fig. 4.16. Volume distilled vs. liquid condensate temperature of used oil at different pressure. Reflux time base is 0.5 s, $R_r = 2$, heat rate = 600 W, cooling temperature = 15 °C.

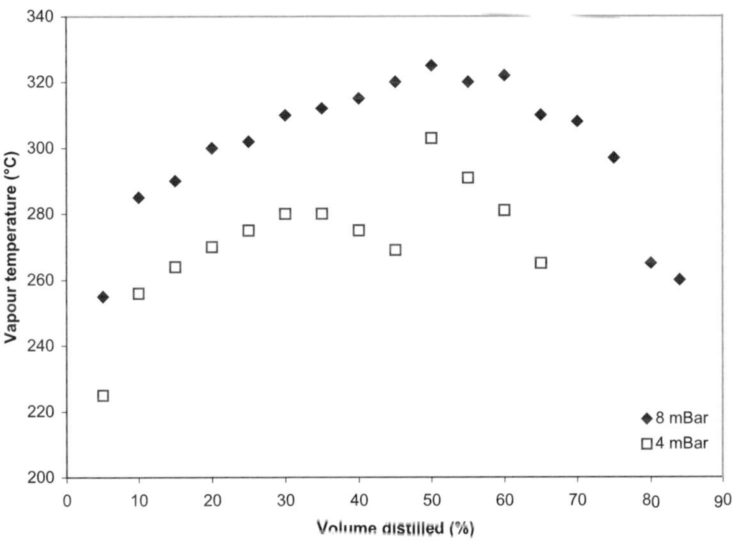

Fig. 4.17. Volume distilled vs. vapor temperature of used oil at different pressure. Reflux time base is 0.5 s, $R_r = 2$, heat rate = 600 W, cooling temperature = 15 °C.

4.4.1.3 *Boiling points.* The behavior and properties of used oil distillation cuts at different boiling points, i.e. 370 °C (the cracking temperature) and 400 °C, were studied as showed in Figs. 4.21–4.26 and Tables 4.12 and 4.13. This study was conducted in order to evaluate the cracking reaction above cracking temperature on the distillate properties. It is

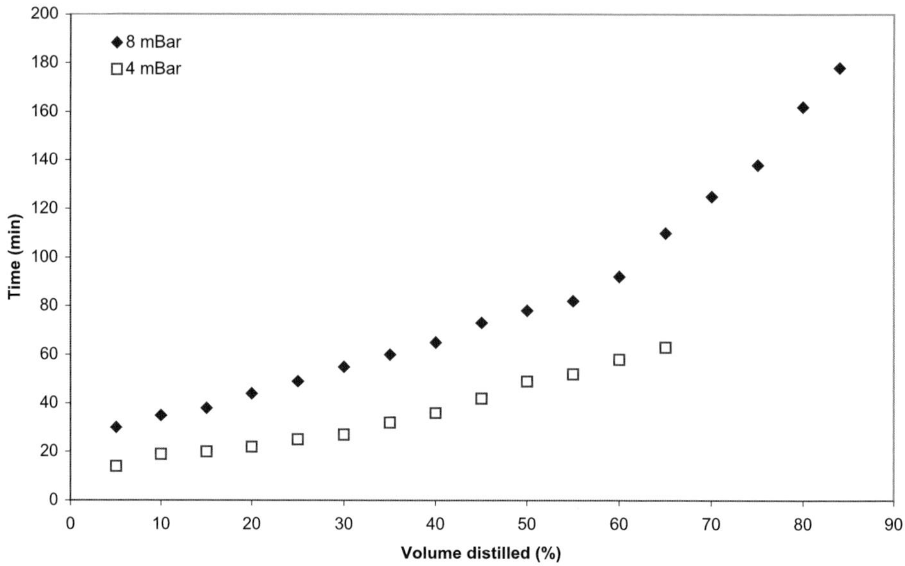

Fig. 4.18. Volume distilled vs. time of used oil at different pressure. Reflux time base is 0.5 s, $R_r = 2$, heat rate = 600 W, cooling temperature = 15 °C.

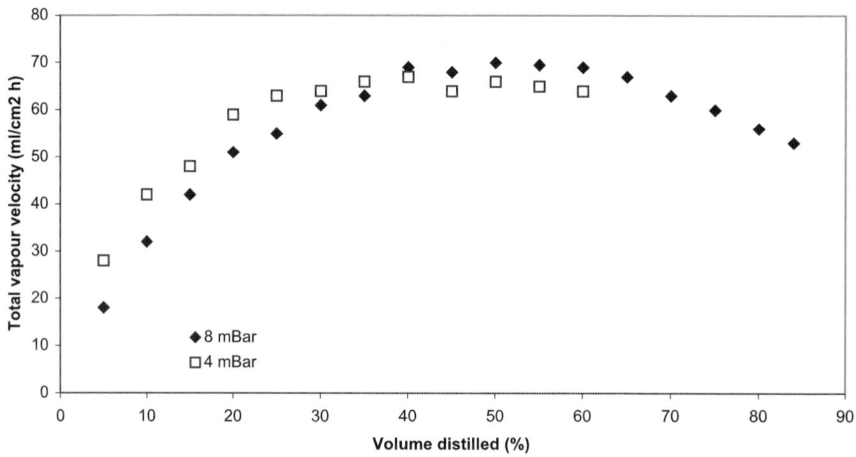

Fig. 4.19. Volume distilled vs. total vapor velocity of used oil at different pressure. Reflux time base is 0.5 s, $R_r = 2$, heat rate = 600 W, cooling temperature = 15 °C.

obvious from Tables 4.8 and 4.12 that the used oil distillation at 370 °C cut point produced less lubricating base oil than the 400 °C cut point. Tables 4.9 and 4.13 indicate that nearly the same distillate properties were obtained for the two different cuts. Cracking reactions have negligible effects on the lubricating oil properties until temperature of 400 °C for the times less than 60 minutes. The distillation stops at 370 °C, even though there is contin-

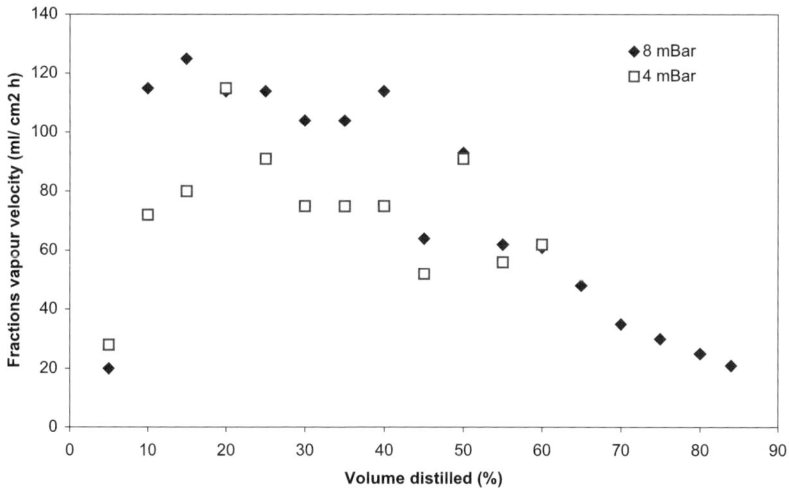

Fig. 4.20. Volume distilled vs. fractions vapor velocity of used oil at different pressure. Reflux time base is 0.5 s, $R_r = 2$, heat rate = 600 W, cooling temperature = 15 °C.

Table 4.10. Fractions obtained for distillation of used oil at 4 mBar pressure

Fraction	Used oil at 4 mBar pressure (vol%)	(wt.%)
Oil base 1	28.6	27.8
Oil base 2	32.8	32.3
Residue	21.4	21.4
Losses	17.1	18.5

Table 4.11. Test analysis for the used oil distillates at 4 mBar pressure

Property	Oil base 1	Oil base 2	Residue
Sp gr	0.885	0.895	0.907
Viscosity (cSt) 40 °C	27.2	–	–
Pour point (°C)	−21	−7	–
Ash (wt.%)	0.02	0.25	1.06
Acid No.	0.37	–	–

uous heat input to the system to maintain constant temperature, i.e. 370 °C, as shown by Fig. 4.21. Liquid condensate temperature and vapor temperature at 370 °C cut point decrease with percentage volume distilled, even though there is continuous heat input, since there is no vaporization that heats the condensed liquid as shown in Figs. 4.22 and 4.23. The general results of the 400 °C cut point gave smooth behavior as shown by the above mentioned figures, which indicate that higher temperature leads to more oil distilled com-

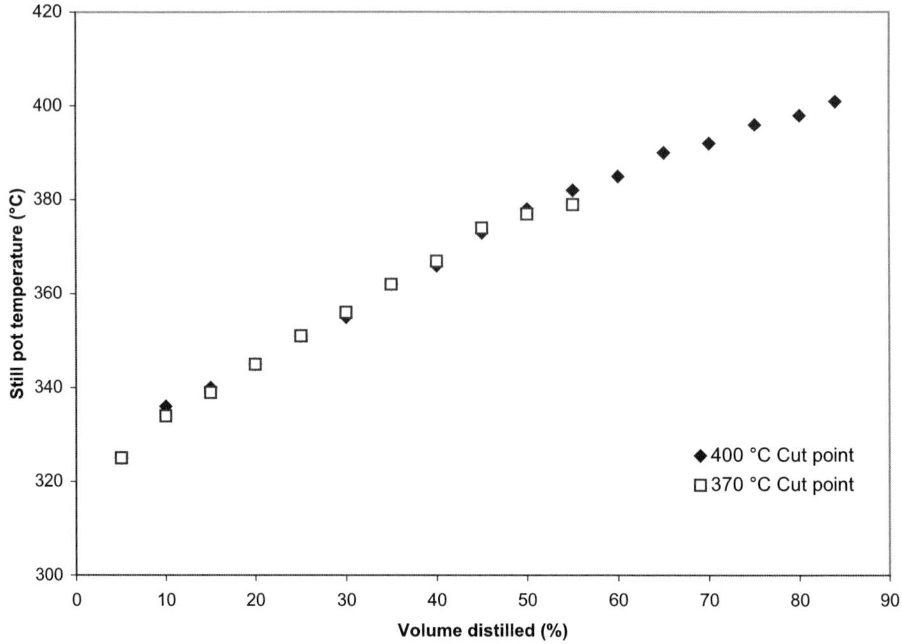

Fig. 4.21. Volume distilled vs. still pot temperature of used oil at different boiling point cuts. Reflux time base is 0.5 s, $R_r = 2$, cooling temperature = 15 °C, $P = 8$ mBar.

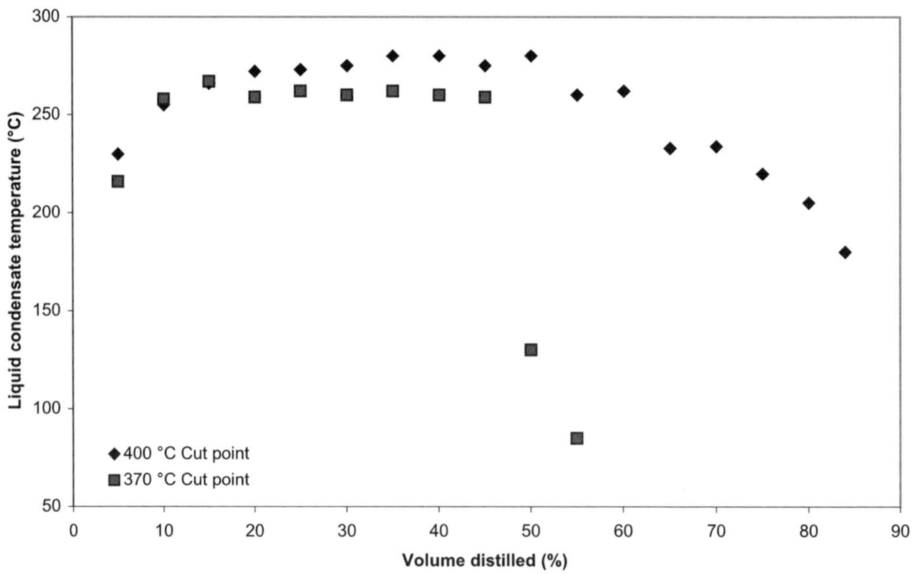

Fig. 4.22. Volume distilled vs. liquid condensate temperature of used oil at different boiling point cuts. Reflux time base is 0.5 s, $R_r = 2$, cooling temperature = 15 °C, $P = 8$ mBar.

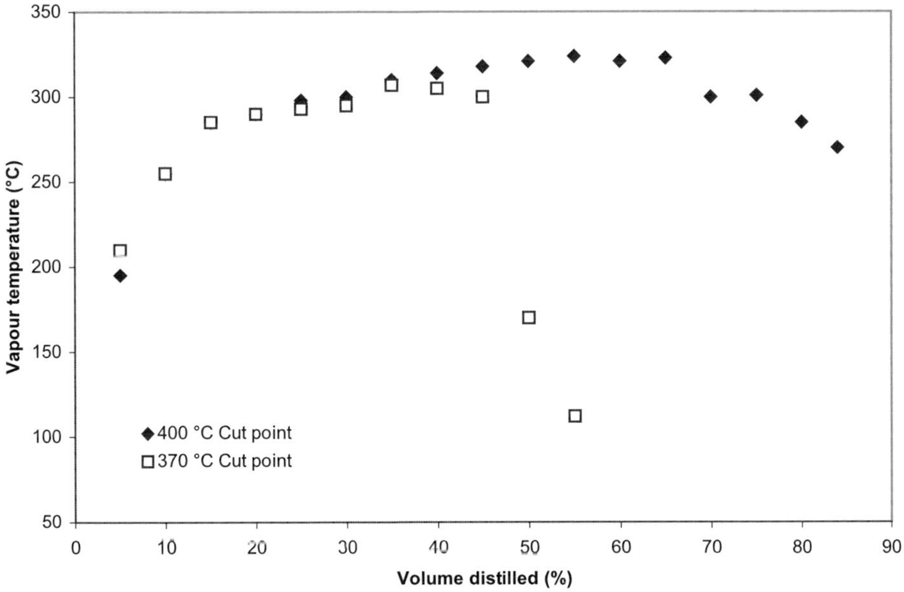

Fig. 4.23. Volume distilled vs. vapor temperature of used oil at different boiling point cuts. Reflux time base is 0.5 s, $R_r = 2$, cooling temperature = 15 °C, $P = 8$ mBar.

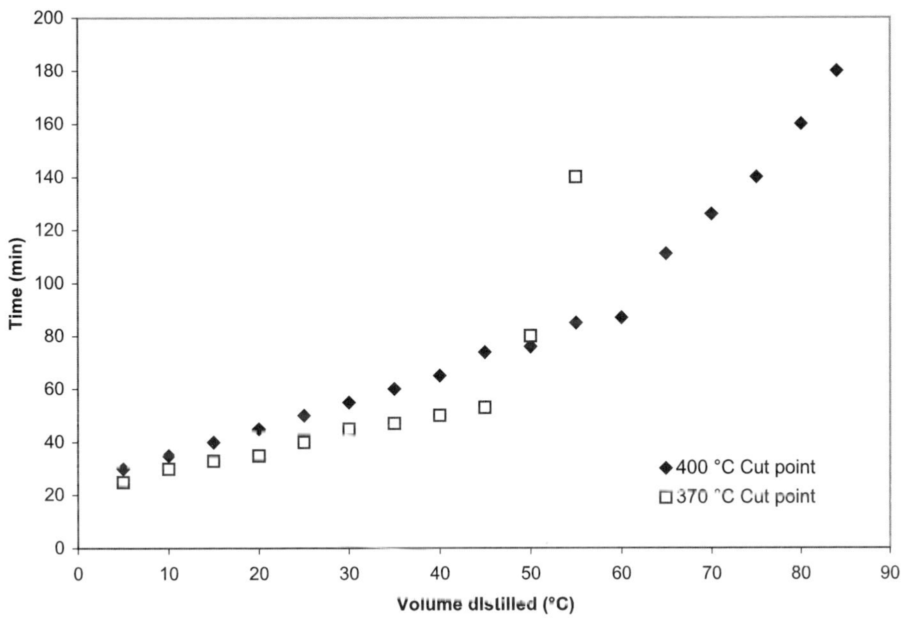

Fig. 4.24. Volume distilled vs. time of used oil at different boiling point cuts. Reflux time base is 0.5 s, $R_r = 2$, cooling temperature = 15 °C, $P = 8$ mBar.

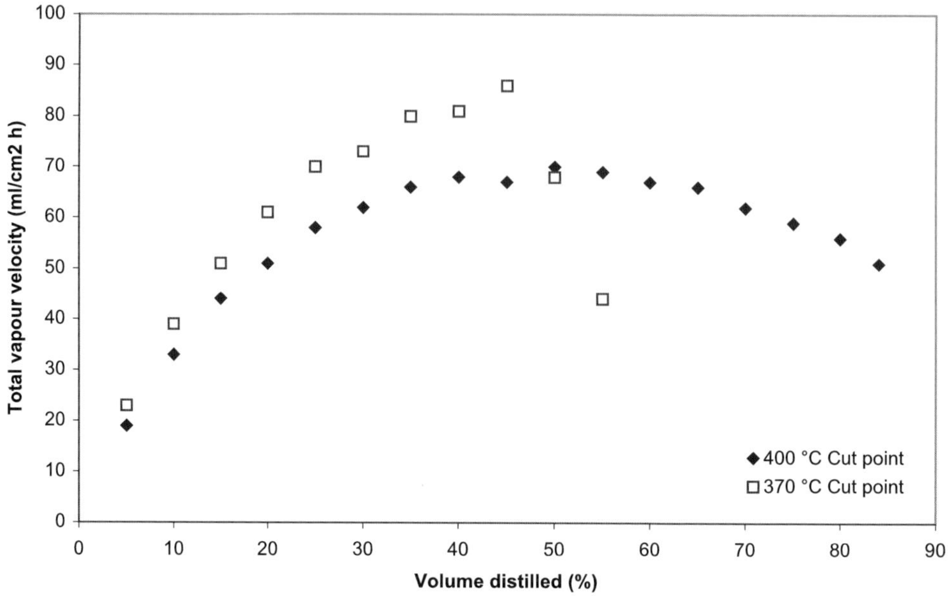

Fig. 4.25. Volume distilled vs. total vapor velocity of used oil at different boiling point cuts. Reflux time base is 0.5 s, $R_r = 2$, cooling temperature $= 15\,^\circ\mathrm{C}$, $P = 8$ mBar.

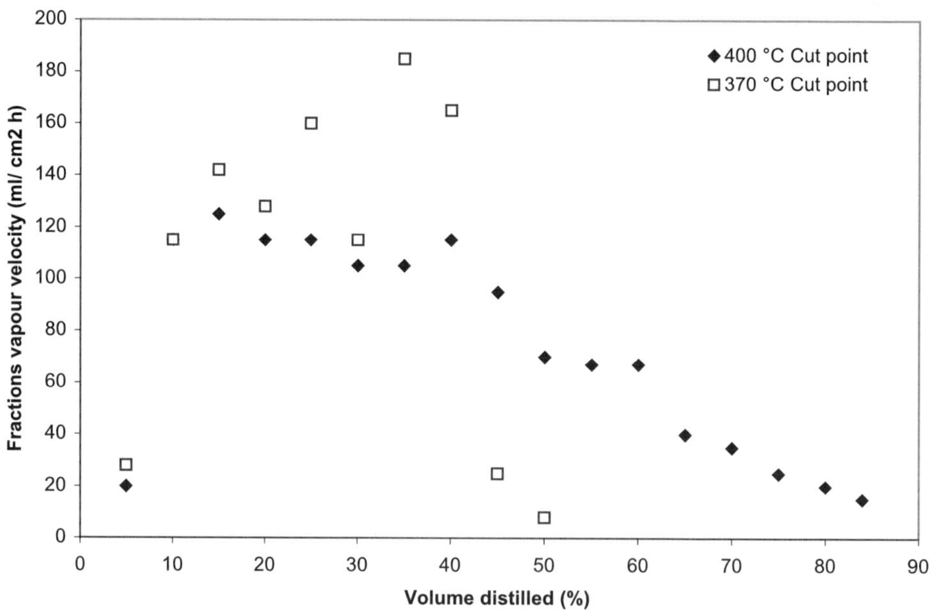

Fig. 4.26. Volume distilled vs. fractions vapor velocity of used oil at different boiling point cuts. Reflux time base is 0.5 s, $R_r = 2$, cooling temperature $= 15\,^\circ\mathrm{C}$, $P = 8$ mBar.

Table 4.12. Fractions obtained for distillation of used oil for a final still pot temperature of 370 °C

Fraction	Used oil (370 °C cut point) (vol%)
Gasoline and water	8.0
Oil base 1	40.0
Oil base 2	14.0
Residue	38.0

Table 4.13. Tests analysis for used oil distillates at a final still pot temperature of 370 °C

Test	Oil base 1 at 370 °C cut point
Viscosity (cSt) 40 °C	40.00
Pour point (°C)	−4.00
Acid No.	0.02

pared with oil distilled at 370 °C cut point. Since it is required to obtain a maximum of oil distillates. The 400 °C temperature was used instead of 370 °C due to the fact that at 400 °C gives 85% of oil distilled compared with 55% of oil distilled at 370 °C. Hence all distillation experiments were conducted at about 400 °C for 8 mBar pressure.

4.4.2 Virgin Lubricating Oil

Experiments on distillation of virgin lubricating oil were carried out by studying the following.

4.4.2.1 *Heat rate.* In order to choose the lowest possible heat rate for the fractionation of virgin oil, two heat rates were examined 400 W and 600 W at constant reflux ratio of 2, pressure of 4 mBar and reflux time base of 0.5 s. The behaviors of virgin oil fractionation under different heat rates are shown in Figs. 4.27–4.32.

The results indicate that 600 W produce satisfactory heat rate for the fractionation process for the specified charge amount of 700 mL. Heat rate does not influence still pot temperature as shown in Fig. 4.27. The volume of oil distilled of 400 W heat rate was about one third of that at the 600 W heat rate. Liquid condensate temperature, vapor temperature, total vapor velocity and fraction vapor velocity are much higher at 600 W heat rate compared with 400 W heat rate as shown by Figs. 4.28–4.32. Since 600 W heat rate gave satisfactory results, higher heating rates were not studied.

4.4.2.2 *Charge.* Different charges were studied namely 600 mL, 700 mL and 1000 mL, in order to choose the lowest volume of charge that produce the best oil cut amounts for testing at constant distillation conditions. The behaviors of virgin oil fractionation under different charges (i.e. 600 mL and 700 mL) are shown in Figs. 4.33–4.38. The charge

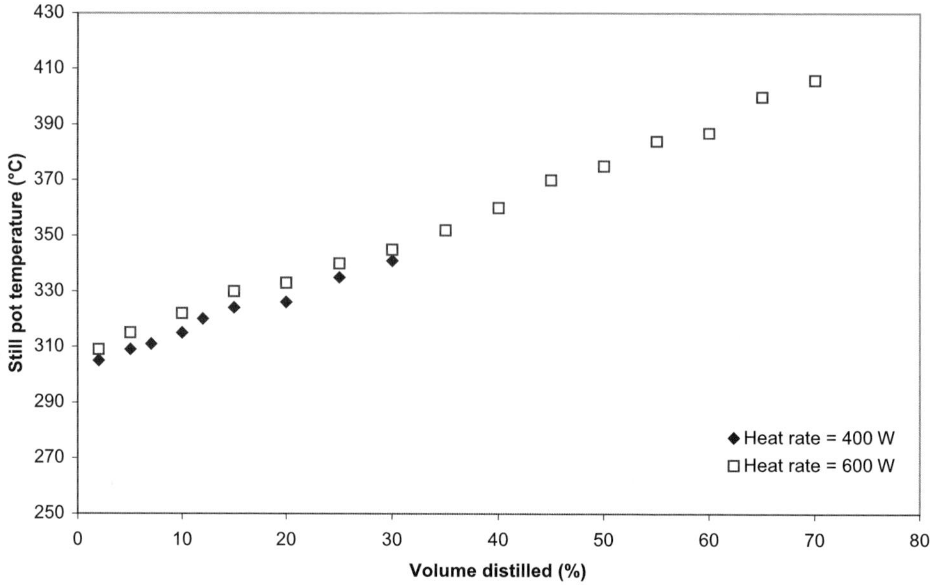

Fig. 4.27. Volume distilled vs. still pot temperature of virgin oil at different heat rates. Reflux time base is 0.5 s, $R_r = 2$, cooling temperature $= 15\,°\mathrm{C}$, $P = 4$ mBar.

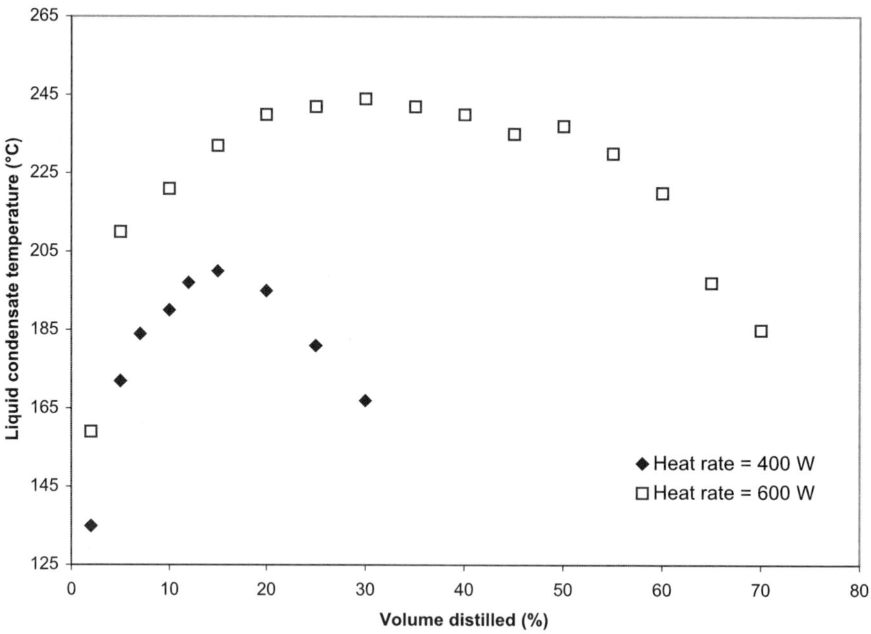

Fig. 4.28. Volume distilled vs. liquid condensate temperature of virgin oil at different heat rates. Reflux time base is 0.5 s, $R_r = 2$, cooling temperature $= 15\,°\mathrm{C}$, $P = 4$ mBar.

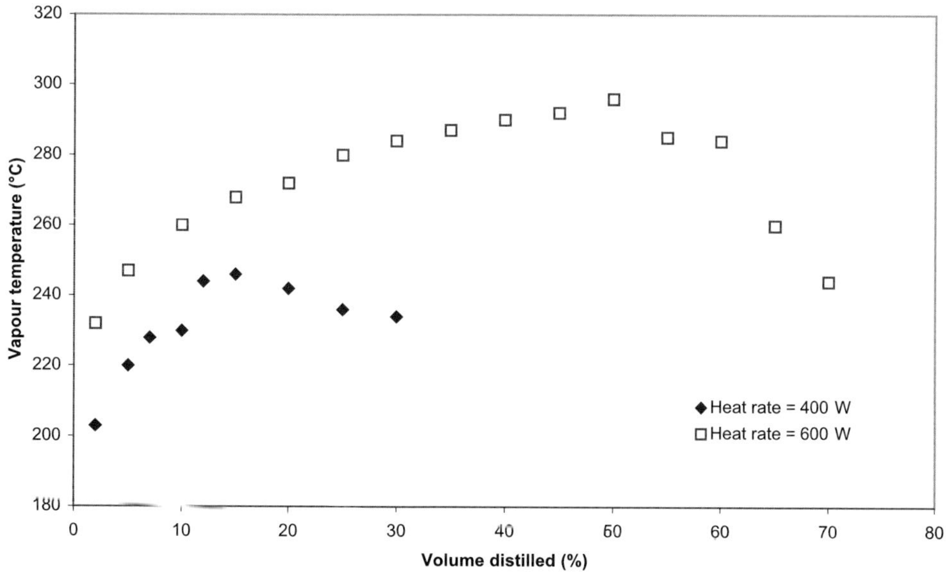

Fig. 4.29. Volume distilled vs. vapor temperature of virgin oil at different heat rates. Reflux time base is 0.5 s, $R_r = 2$, cooling temperature = 15 °C, $P = 4$ mBar.

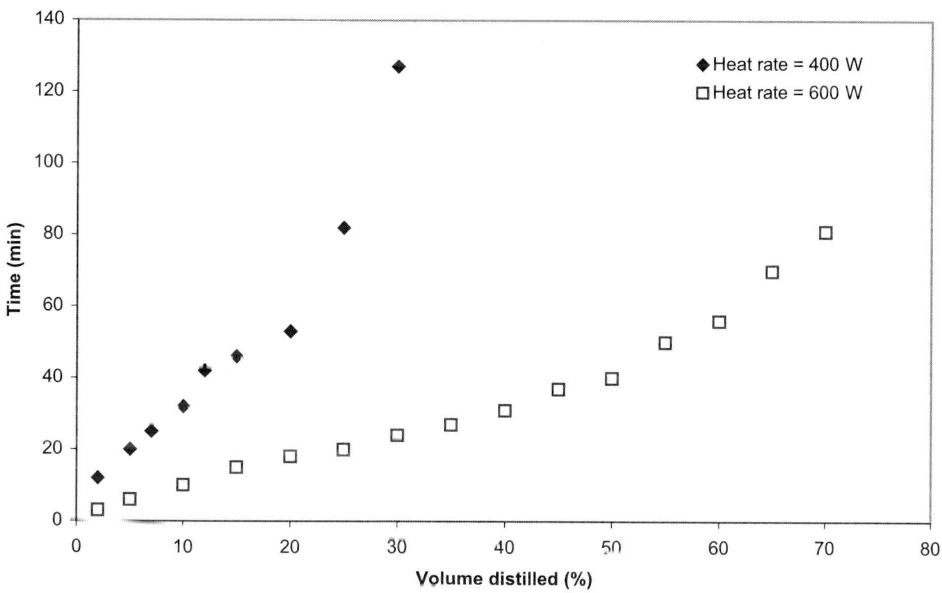

Fig. 4.30. Volume distilled vs. time of virgin oil at different heat rates. Reflux time base is 0.5 s, $R_r = 2$, cooling temperature = 15 °C, $P = 4$ mBar.

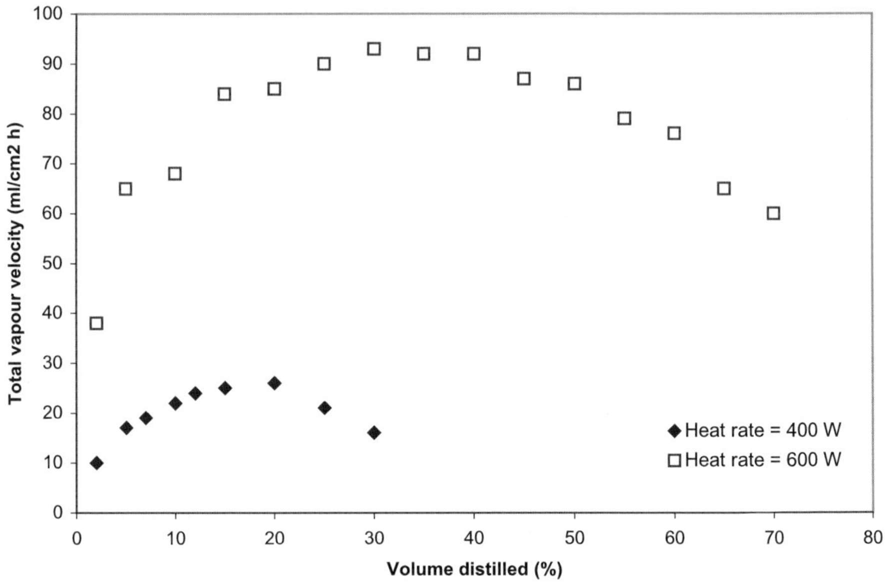

Fig. 4.31. Volume distilled vs. total vapor velocity of virgin oil at different heat rates. Reflux time base is 0.5 s, $R_r = 2$, cooling temperature = 15 °C, $P = 4$ mBar.

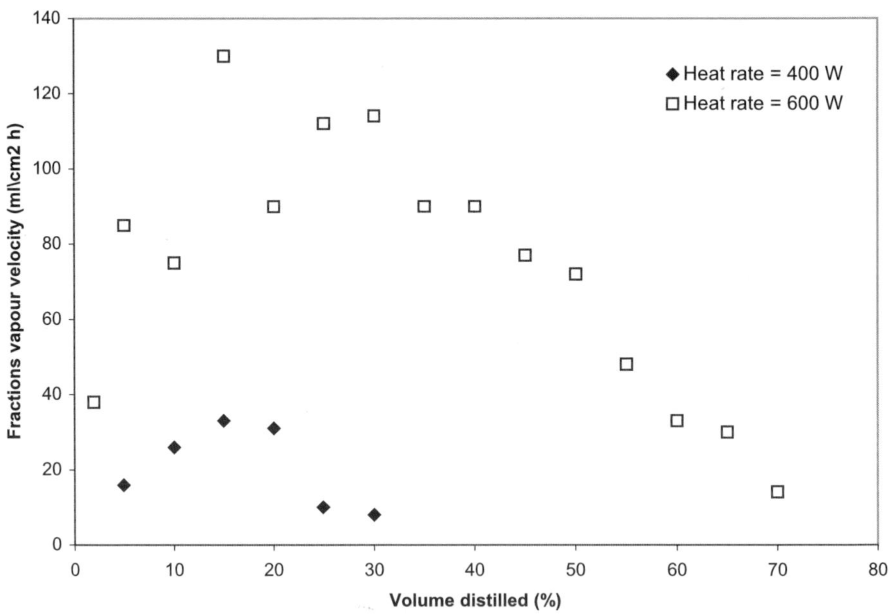

Fig. 4.32. Volume distilled vs. fractions vapor velocity of virgin oil at different heat rates. Reflux time base is 0.5 s, $R_r = 2$, cooling temperature = 15 °C, $P = 4$ mBar.

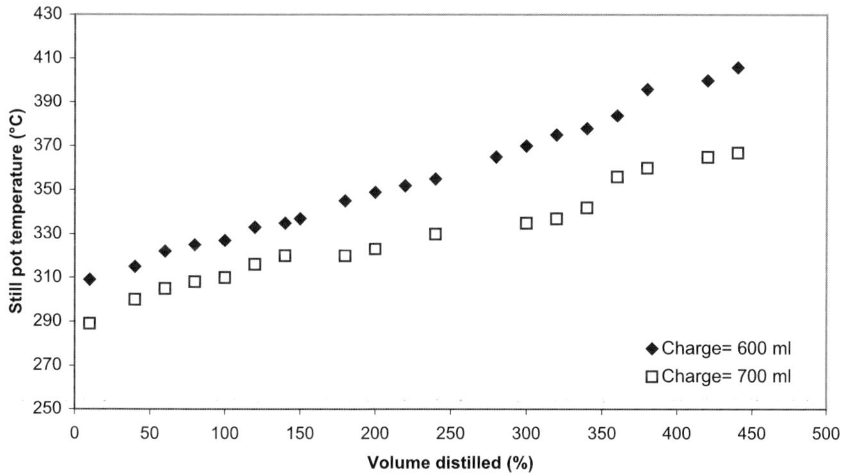

Fig. 4.33. Volume distilled vs. still pot temperature of virgin oil at different charge amount. Reflux time base is 0.5 s, $R_r = 2$, cooling temperature = 15 °C, $P = 4$ mBar, heat rate = 600 W.

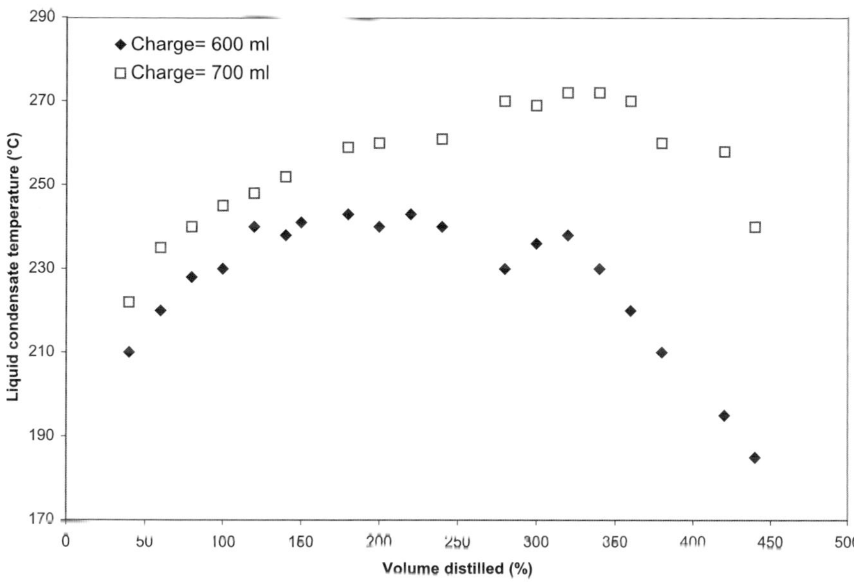

Fig. 4.34. Volume distilled vs. liquid condensate temperature of virgin oil at different charge amount. Reflux time base is 0.5 s, $R_r = 2$, cooling temperature = 15 °C, $P = 4$ mBar, heat rate = 600 W.

of 700 mL was found suitable for the distillation apparatus. Unsteady state results were obtained when using 600 mL charge compared with 700 mL charge, which gave steady state behavior similar to that of 1000 mL charges. The unsteady state condition is due

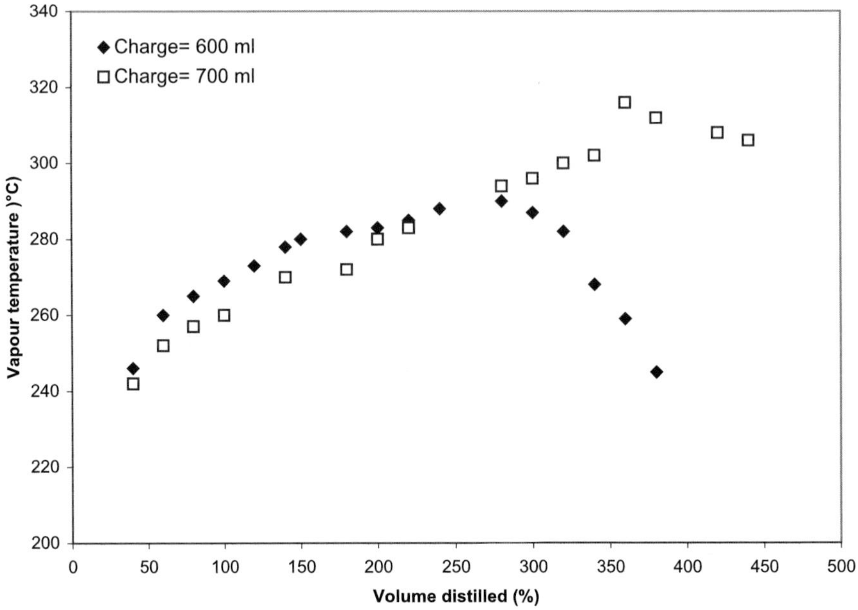

Fig. 4.35. Volume distilled vs. vapor temperature of virgin oil at different charge amount. Reflux time base is 0.5 s, $R_r = 2$, cooling temperature = 15 °C, $P = 4$ mBar, heat rate = 600 W.

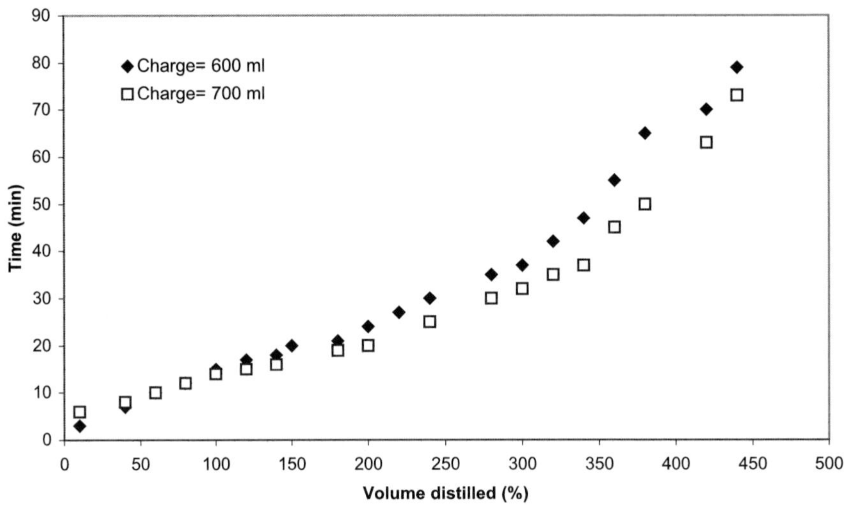

Fig. 4.36. Volume distilled vs. time of virgin oil at different charge amount. Reflux time base is 0.5 s, $R_r = 2$, cooling temperature = 15 °C, $P = 4$ mBar, heat rate = 600 W.

to possible partial condensation of vapor in the distillation column as indicated by visual observation during operation. 700 mL charge was chosen to be the representing charge.

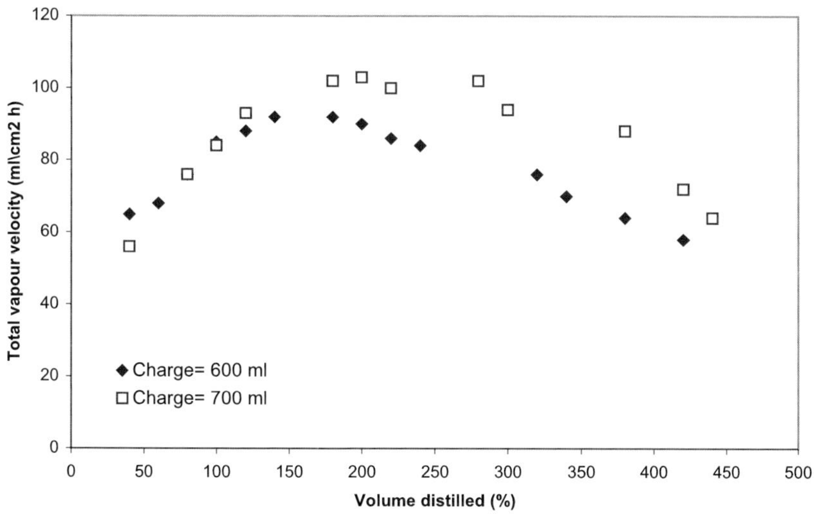

Fig. 4.37. Volume distilled vs. total vapor velocity of virgin oil at different charge amount. Reflux time base is 0.5 s, $R_r = 2$, cooling temperature = 15 °C, $P = 4$ mBar, heat rate = 600 W.

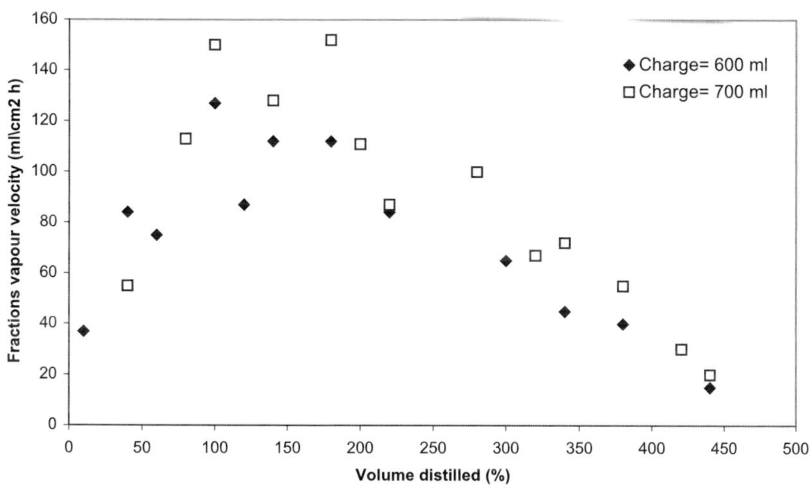

Fig. 4.38. Volume distilled vs. fractions vapor velocity of virgin oil at different charge amount. Reflux time base is 0.5 s, $R_r = 2$, cooling temperature = 15 °C, $P = 4$ mBar, heat rate = 600 W.

4.4.3 Solvent Treated Oil

Experiments on fractionation of solvent treated oils were carried out by studying the following.

4.4.3.1 *Reflux ratio.* The effects of reflux ratio, i.e. 0.5, 2, 3, 4, and 6 on the fractionation behaviors of solvent treated oil and its distillate properties were investigated at a reflux

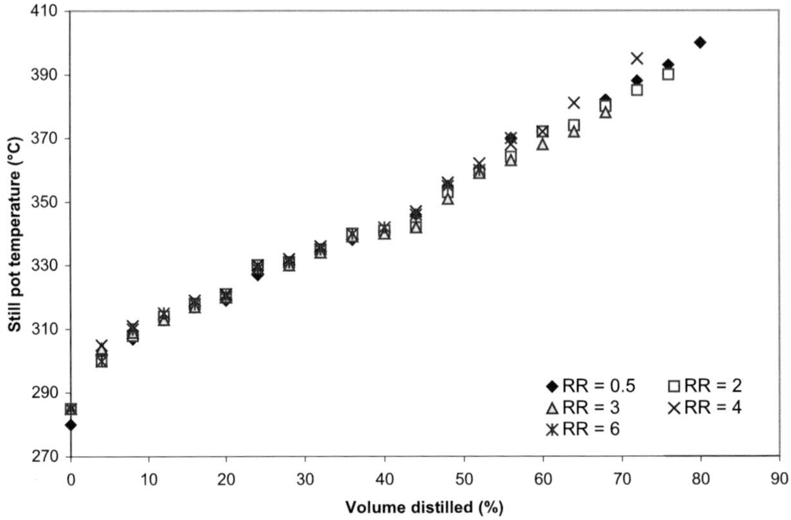

Fig. 4.39. Volume distilled vs. still pot temperature of solvent treated oil at different reflux ratio, reflux time base of 5 s, $P = 4$ mBar, heat rate = 600 W, cooling temperature = 15 °C and a 700 mL charge.

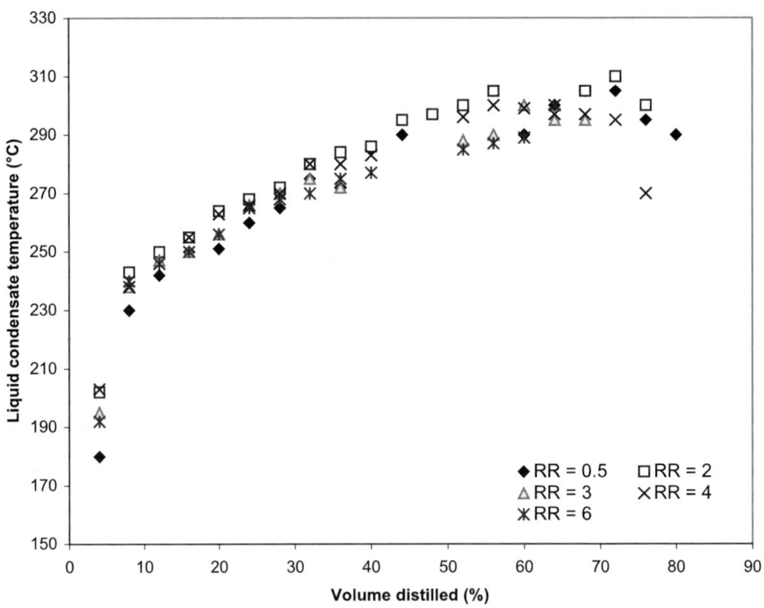

Fig. 4.40. Percent volume distilled vs. liquid condensate temperature of solvent treated oil at different reflux ratio, reflux time base of 5 s, $P = 4$ mBar, heat rate = 600 W, cooling temperature = 15 °C and a 700 mL charge.

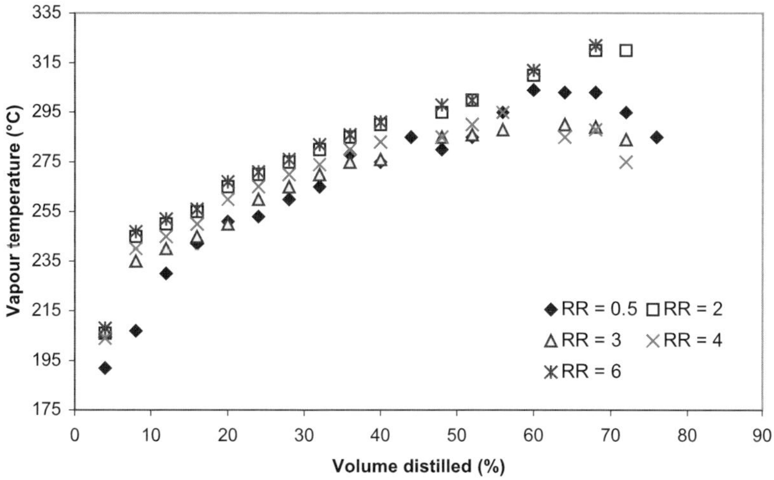

Fig. 4.41. Volume distilled vs. vapor temperature of solvent treated oil at different reflux ratio, reflux time base of 5 s, $P = 4$ mBar, heat rate = 600 W, cooling temperature = 15 °C and a 700 mL charge.

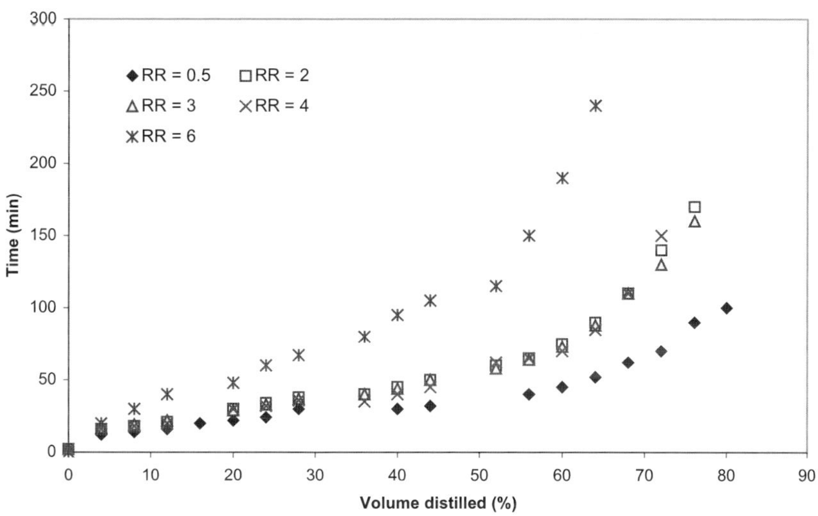

Fig. 4.42. Volume distilled vs. time of solvent treated oil at different reflux ratio, reflux time base of 5 s, $P = 4$ mBar, heat rate = 600 W, cooling temperature = 15 °C and a 700 mL charge.

time base of 5 s as shown in Figs. 4.39–4.47 and Tables 4.14–4.17. Similarly the effects of reflux ratio, i.e. 0.5 and 2 at a reflux time base of 0.5 s on the fractionation behavior of solvent treated oil and its distillate properties were studied as shown in Figs. 4.48–4.53 and Tables 4.18 and 4.19.

Table 4.14 indicates that almost the same amounts of oil cuts were produced at different reflux ratios except at the reflux ratio of 0.5, which produces a higher amount. All the re-

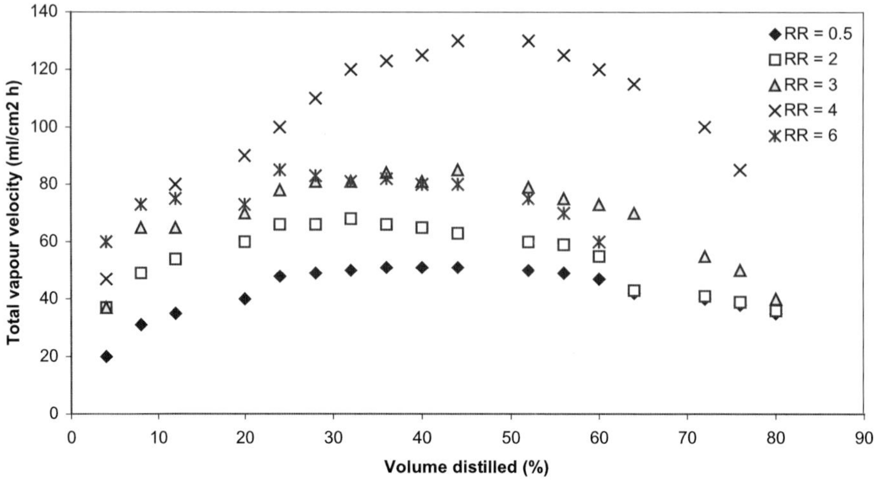

Fig. 4.43. Percent volume distilled vs. total vapor velocity of solvent treated oil at different reflux ratio, reflux time base of 5 s, $P = 4$ mBar, heat rate = 600 W, cooling temperature = 15 °C and a 700 mL charge.

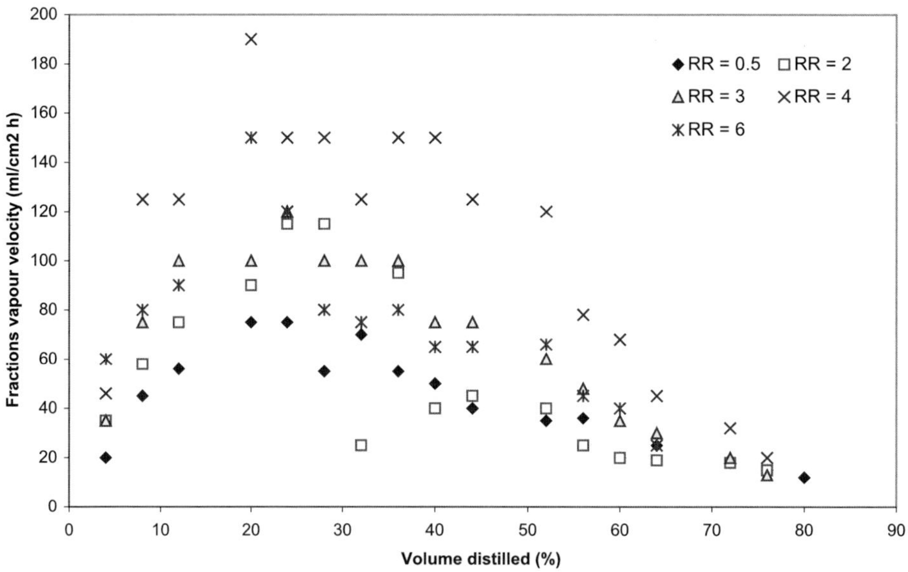

Fig. 4.44. Volume distilled vs. fractions vapor velocity of solvent treated oil at different reflux ratio, reflux time base of 5 s, $P = 4$ mBar, heat rate = 600 W, cooling temperature = 15 °C and a 700 mL charge.

flux ratio operations showed nearly the same amount of lost material. Tables 4.15 and 4.16 show that for base oil cut properties, that all reflux ratios produce nearly the same values of Sp gr, viscosity, flash point, and pour point. That means that changing the reflux ratio does not affect these oil properties. Ash content and acid number were significantly improved

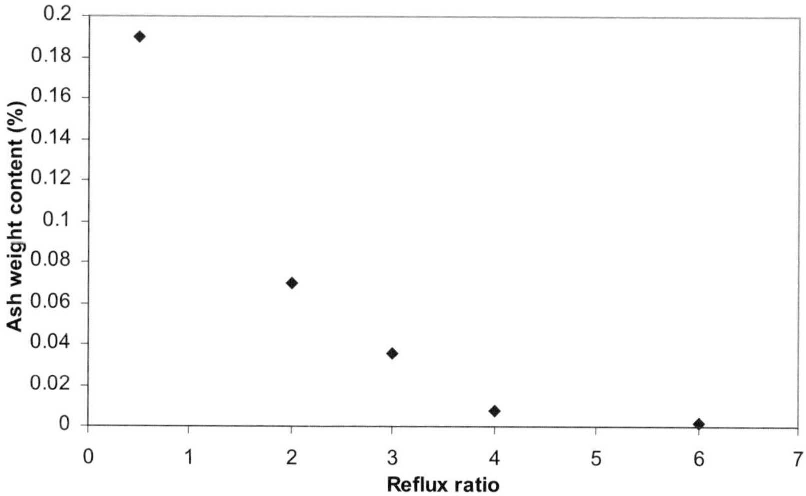

Fig. 4.45. Reflux ratio vs. ash content for vacuum distillation of solvent treated oil at 4 mBar.

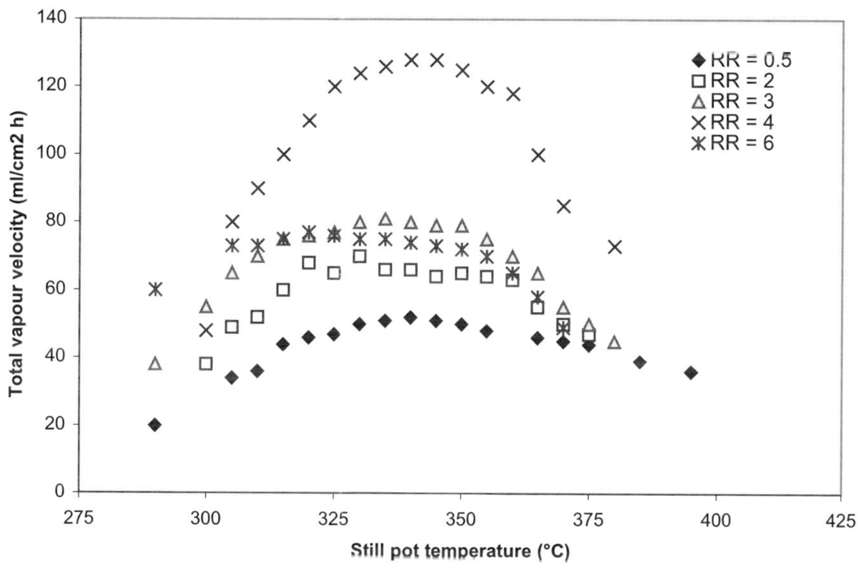

Fig. 4.46. Still pot temperature vs. total vapor velocity of solvent treated oil at different reflux ratio, $P = 4$ mBar, heat rate = 600 W, cooling temperature = 15 °C and a 700 mL charge.

when reflux ratio increased. All oil distillates are suitable for reuse except at the reflux ratio of 0.5. Table 4.17 indicates that high Sp gr and ash content of residue make the residue impossible to use as lubricating oil. Still pot temperature, liquid condensate temperature and vapor temperature are not influenced by changing the reflux ratio as shown in Figs. 4.39–4.41 under constant conditions. There are slight variations of temperature measurements

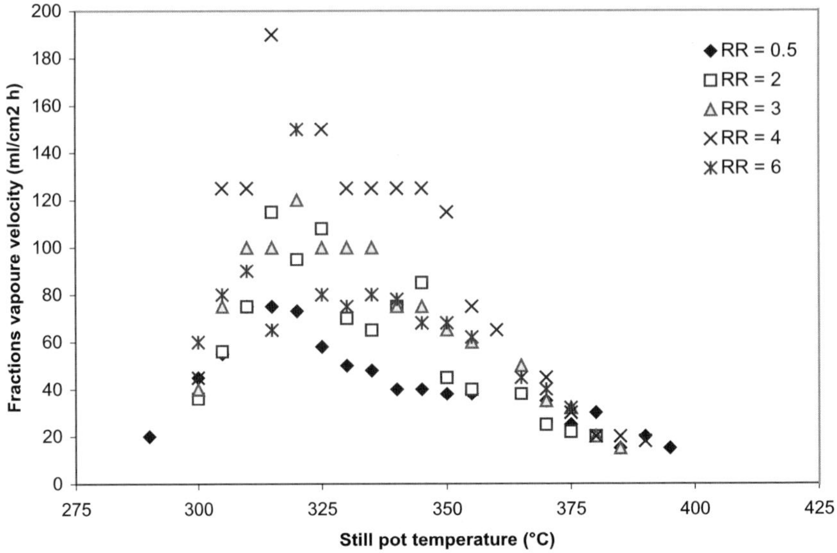

Fig. 4.47. Still pot temperature vs. fractions vapor velocity of solvent treated oil at different reflux ratio, $P = 4$ mBar, heat rate = 600 W, cooling temperature = 15 °C and a 700 mL charge.

Table 4.14. Fractions obtained for distillation of solvent treated oil at different reflux ratios with reflux time base of 5 s and 4 mBar pressure

Fraction	Reflux ratio				
	0.5	2	3	4	6
Oil base 1	48.9	50.2	46.9	47.9	48.0
Oil base 2	32.3	25.5	22.7	26.4	18.4
Residue	14.5	20.9	26.2	21.1	28.8
Loss	4.3	3.4	4.2	4.6	4.8

Table 4.15. Tests analysis for oil base 1 fraction distilled for solvent treated oil at different reflux ratio at reflux time of 5 s and 4 mBar pressure

Test	Reflux ratio				
	0.5	2	3	4	6
Sp gr	0.887	0.888	0.888	0.889	0.888
Viscosity (cSt) 40 °C	37.4	40.5	39.2	37.9	37.0
Viscosity (cSt) 100 °C			6.9		
Flash point (°C)	190		196	195	
Pour point (°C)	−11	−8	−6	−7	−7
Ash (wt.%)	0.19	0.07	0.03	0.01	>0.01
Acid No.	0.40	0.20	0.32	0.18	0.04
Copper corrosion					La
Viscosity index			130		

Table 4.16. Tests analysis for oil base 2 fractions distilled for solvent treated oil at different reflux ratio at reflux time of 5 s and 4 mBar pressure

Test	Reflux ratio				
	0.5	2	3	4	6
Sp gr	0.899	0.897	0.897	0.897	0.897
Viscosity (cSt) 40 °C	44.6	46.2	47.6	45.7	41.2
Viscosity (cSt) 100 °C	10.8				9.5
Flash point (°C)	203				200
Pour point (°C)	−8	−5	−4	−4	−2
Ash (wt.%)			0.18		0.02
Acid No.					0.26
Copper corrosion					La
Viscosity index	163				161

Table 4.17. Tests analysis for residue fraction distilled for solvent treated oil at different reflux ratio at reflux time of 5 s and 4 mBar pressure

Test	Reflux ratio				
	0.5	2	3	4	6
Sp gr		0.902			
Ash (wt.%)	0.91	0.58	0.12	0.05	0.02

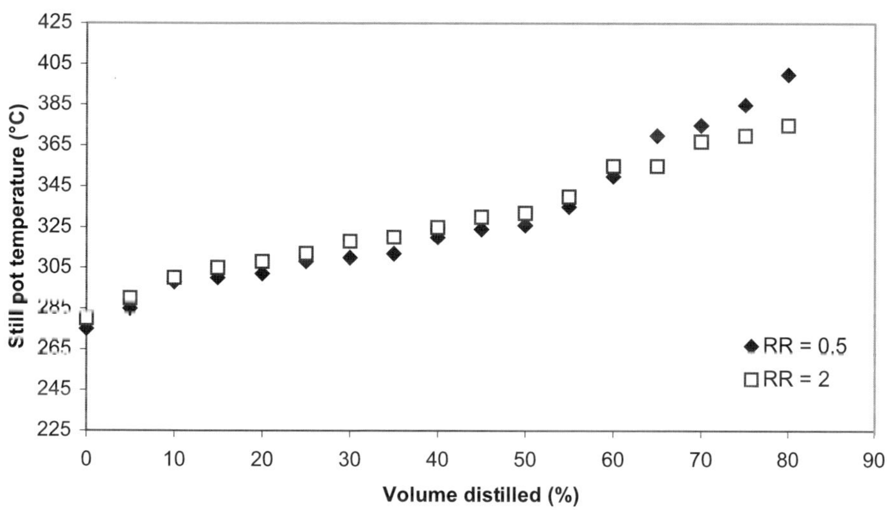

Fig. 4.48. Percent volume distilled vs. still pot temperature of solvent treated oil at different reflux ratio, reflux time base of 0.5 s, $P = 4$ mBar, heat rate = 600 W and cooling temperature = 15 °C.

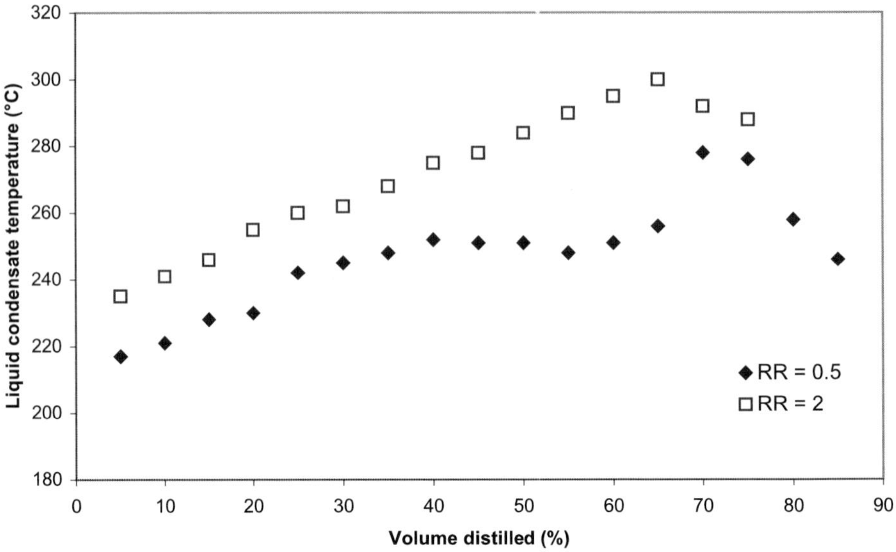

Fig. 4.49. Percent volume distilled vs. liquid condensate temperature of solvent treated oil at different reflux ratio, reflux time base of 0.5 s, $P = 4$ mBar, heat rate = 600 W and cooling temperature = 15 °C.

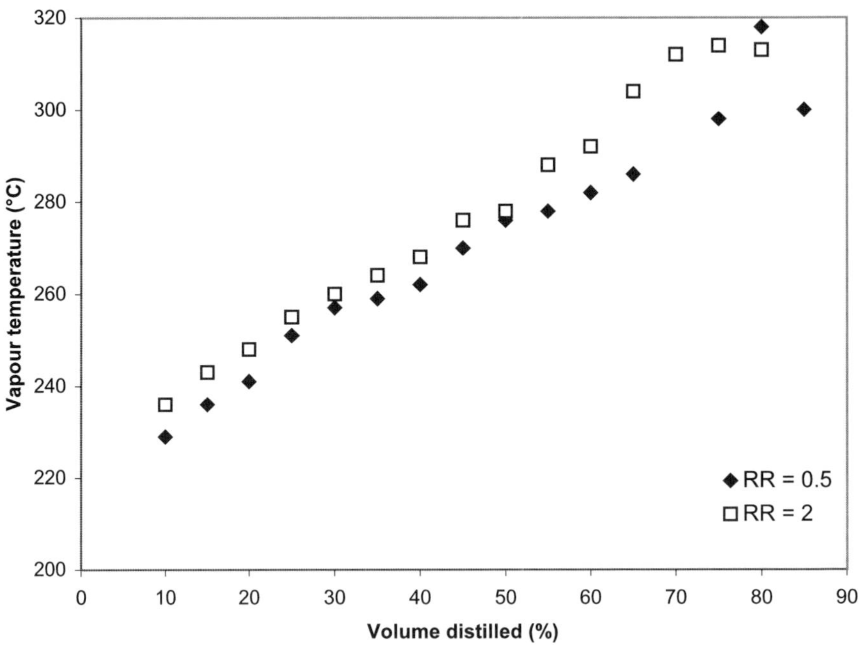

Fig. 4.50. Percent volume distilled vs. vapor temperature of solvent treated oil at different reflux ratio, reflux time base of 0.5 s, $P = 4$ mBar, heat rate = 600 W and cooling temperature = 15 °C.

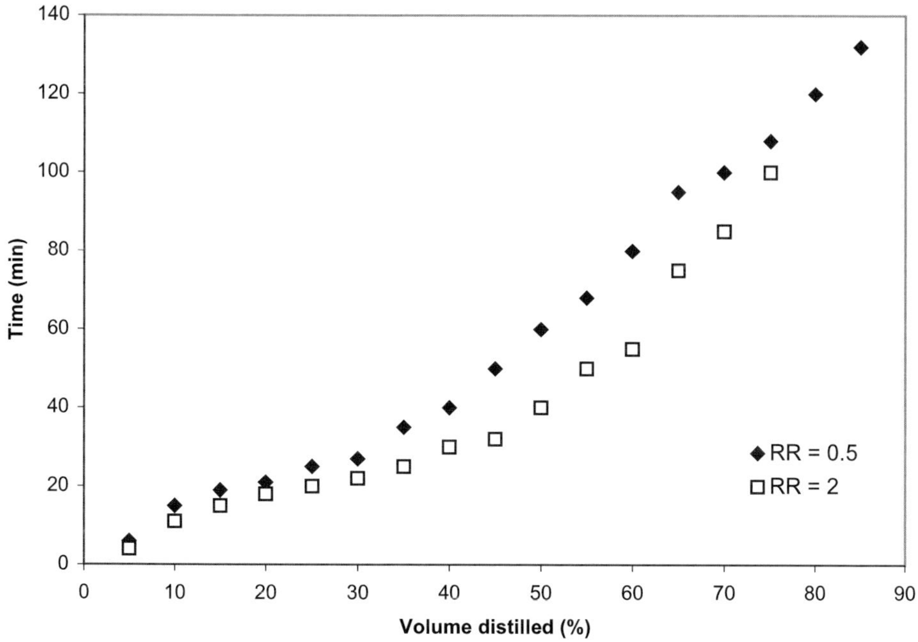

Fig. 4.51. Percent volume distilled vs. time of solvent treated oil at different reflux ratio, reflux time base of 0.5 s, $P = 4$ mBar, heat rate = 600 W and cooling temperature = 15 °C.

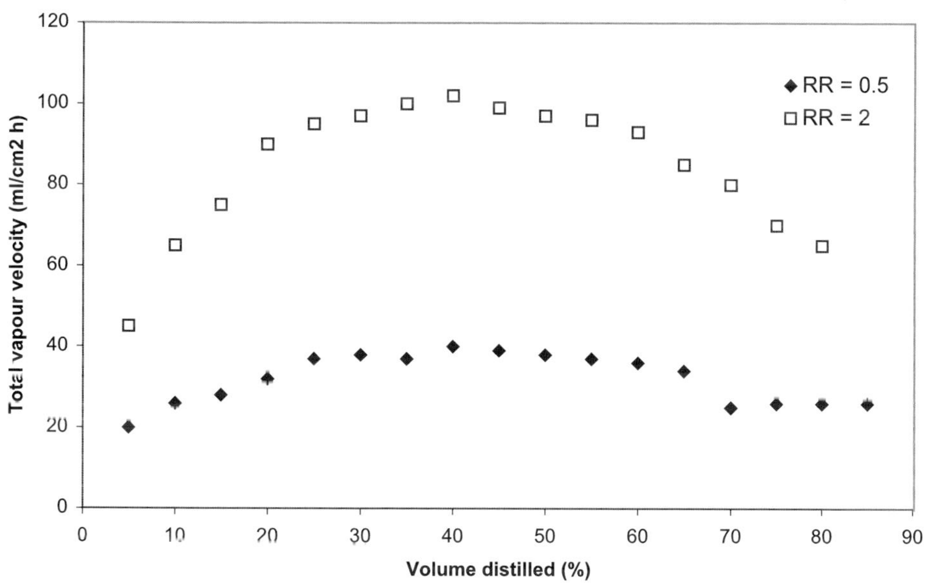

Fig. 4.52. Percent volume distilled vs. total vapor velocity of solvent treated oil at different reflux ratio, reflux time base of 0.5 s, $P = 4$ mBar, heat rate = 600 W and cooling temperature = 15 °C.

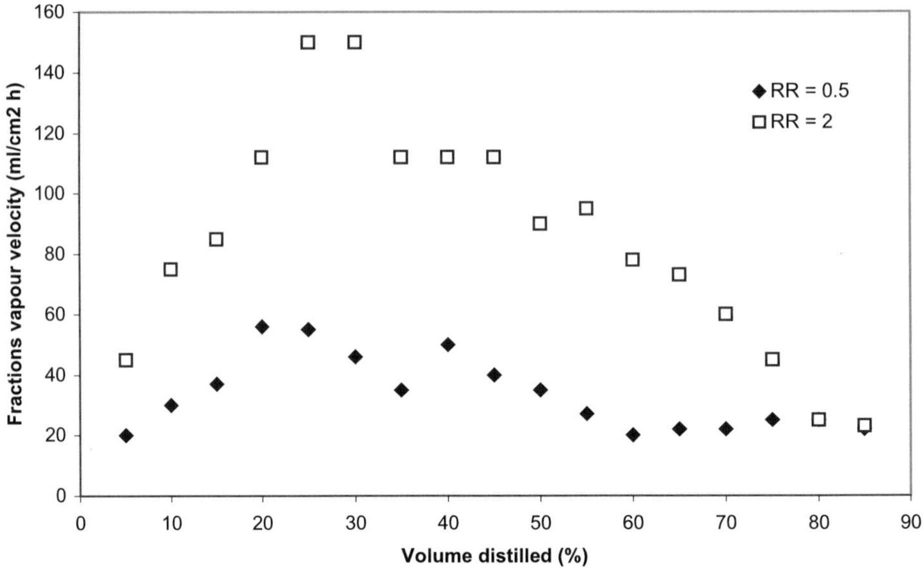

Fig. 4.53. Percent volume distilled vs. fraction vapor velocity of solvent treated oil at different reflux ratio, reflux time base of 0.5 s, $P = 4$ mBar, heat rate = 600 W and cooling temperature = 15 °C.

Table 4.18. Fractions obtained for distillation of solvent treated oil at different reflux ratios with reflux time base of 0.5 s and 4 mBar pressure

Fraction	Reflux ratio	
	0.5	2
Oil base 1	48.4	49.4
Oil base 2	33.50	25.3
Residue	16.5	21.1
Loss	1.6	4.2

Table 4.19. Tests analysis for fractions distilled for solvent treated oil at different reflux ratio at reflux time of 0.5 s and 4 mBar pressure

Test	Reflux ratio = 0.5		Reflux ratio = 2	
	Oil base 1	Oil base 2	Oil base 1	Oil base 2
Sp gr	0.887	0.897	0.888	0.899
Viscosity (cSt) 40 °C	42.8		37.1	
Pour point (°C)	−8		−10	−10
Ash (wt.%)	>0.01	>0.01	0.01	
Acid No.			0.31	

that cannot be evaluated accurately. The only effect that can be seen is in the time of distillation, since the larger refluxes need longer time to be re-distilled as shown in Fig. 4.42, while the temperature readings were not influenced by the reflux ratio. Total vapor velocity against percentage volume distilled for different reflux ratios shows different variation as shown in Fig. 4.43. The total vapor velocity decreases with reflux ratio in the following order 4, 6, 3, 2, 0.5. A reflux ratio of 4 is the most successful ratio due to the fact that reflux ratio of 6 gave a higher quantity of vapor load in longer time of distillation, while reflux ratios of 0.5, 2 and 3 gave a lower quantity of vapor load, but shorter time. Reflux ratio of 4 gave a lower quantity of vapor load than reflux ratio of 6, but at more moderate times.

At a reflux time base of 0.5 s, Table 4.18 shows that larger base oil cuts were obtained for reflux ratio of 0.5 than reflux ratio of 2. Table 4.19 shows that the reflux ratio does not affect the properties of Sp gr, viscosity and pour point for the distillates, while the ash content for the distillates was very low. The two oil cuts are suitable for re-use.

It was found that the reflux time base of 0.5 s gave unsatisfactory results with a reflux ratio of 0.5 due to unsteady state conditions in the distillation column, where vapor/liquid equilibrium exists. This gave unreasonable behavior as shown in Figs. 4.48–4.53. A reflux ratio of 2 in the case of a reflux time base of 5 s gave similar behavior compared with a reflux ratio of 2 and base time of 0.5 s. A reflux ratio of 0.5 for a reflux time base of 5 s gave different behavior compared with a reflux ratio of 0.5 for a reflux time base of 0.5 s. The above indicates that the unsteady state conditions depend on the reflux time base and not on the reflux ratio.

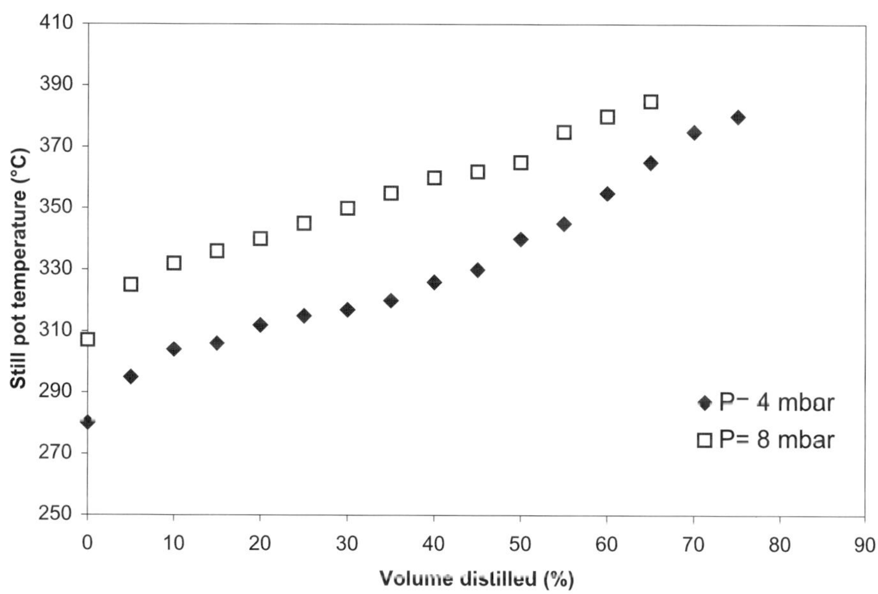

Fig. 4.54 Volume distilled vs. still pot temperature for solvent treated oil at different pressures, $R_r = 2$, heat rate = 600 W, cooling temperature = 15 °C, charge = 700 mL and reflux time base = 0.5 s.

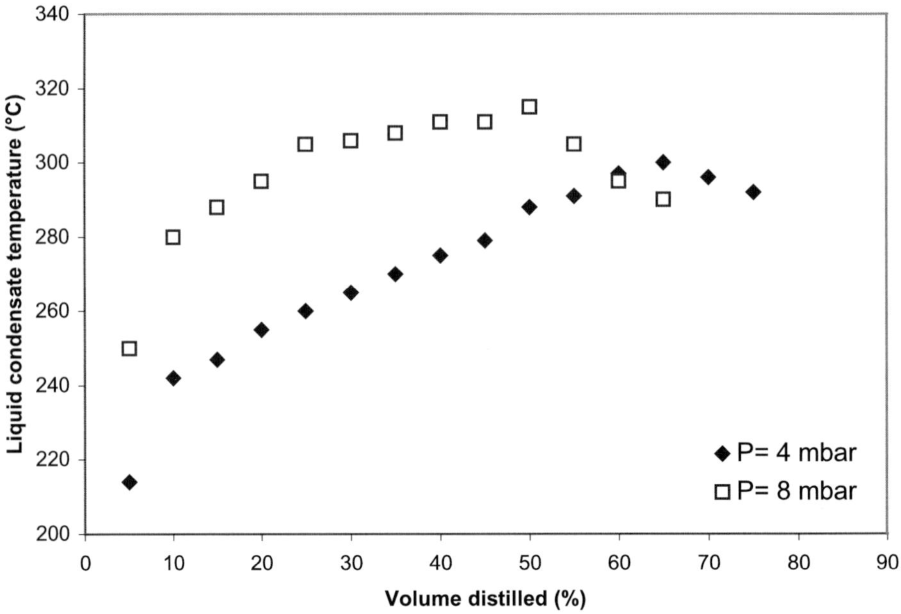

Fig. 4.55. Volume distilled vs. liquid condensate temperature for solvent treated oil at different pressures, $R_r = 2$, heat rate = 600 W, cooling temperature = 15 °C, charge = 700 mL and reflux time base = 0.5 s.

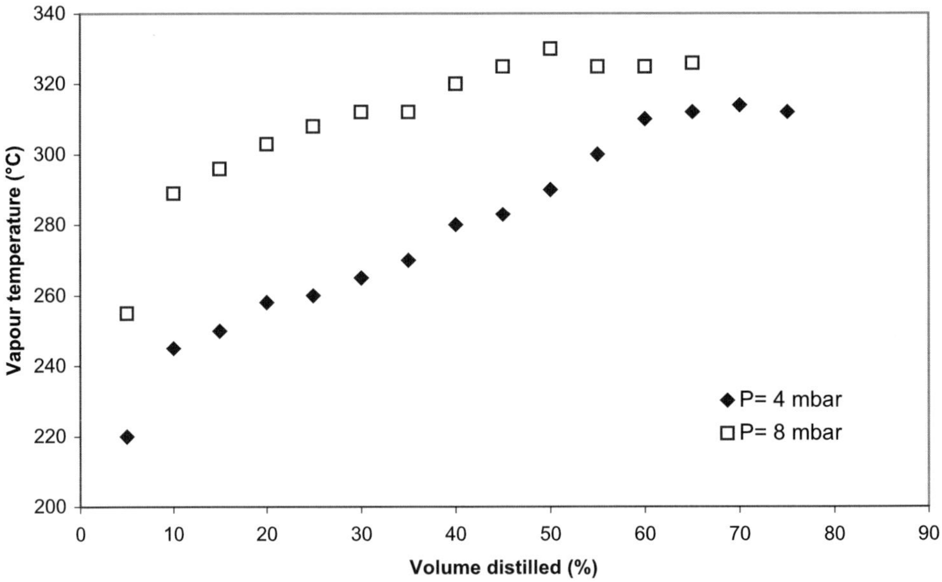

Fig. 4.56. Volume distilled vs. vapor temperature for solvent treated oil at different pressures, $R_r = 2$, heat rate = 600 W, cooling temperature = 15 °C, charge = 700 mL and reflux time base = 0.5 s.

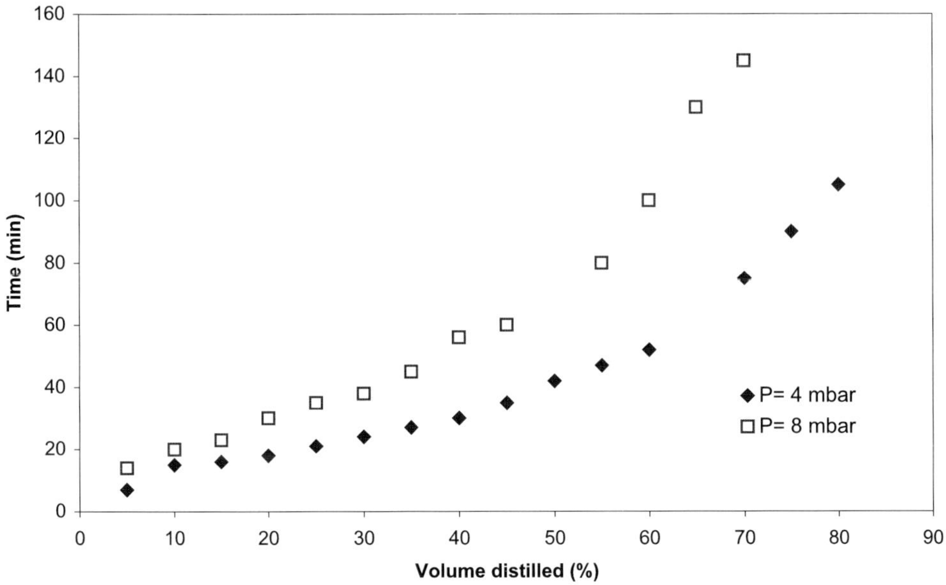

Fig. 4.57. Volume distilled vs. time for solvent treated oil at different pressures, $R_r = 2$, heat rate = 600 W, cooling temperature = 15 °C, charge = 700 mL and reflux time base = 0.5 s.

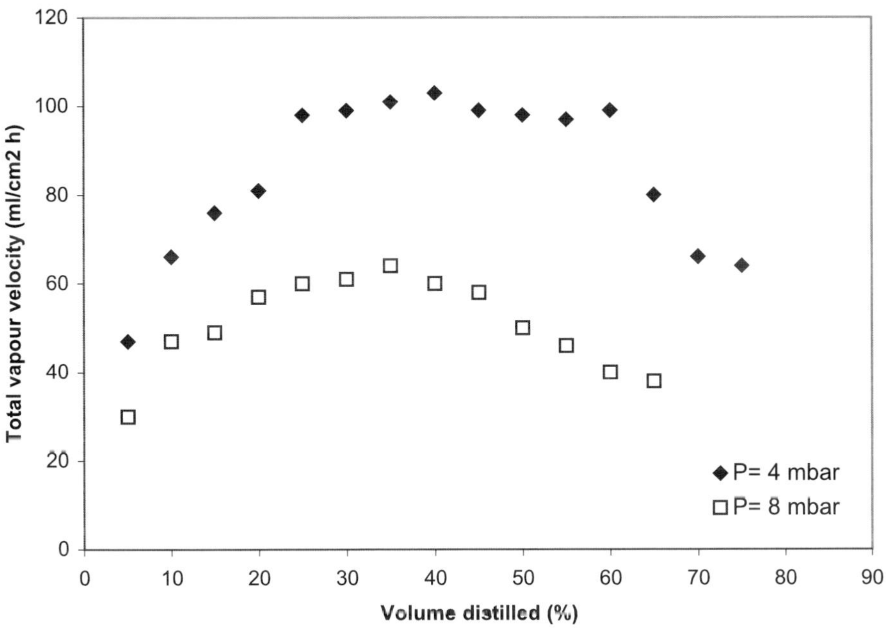

Fig. 4.58. Volume distilled vs. total vapor velocity for solvent treated oil at different pressures, $R_r = 2$, heat rate = 600 W, cooling temperature = 15 °C, charge = 700 mL and reflux time base = 0.5 s.

4.4.3.2 *Vacuum pressure.* The fractionation behaviors of solvent treated oil at different vacuum pressures, i.e. 4 mBar and 8 mBar, were studied as shown in Figs. 4.48–4.59 and Table 4.20. Tables 4.19 and 4.20 indicate that no major differences were noticed in distillate properties. Figures 4.48–4.59 indicate that 4 mBar pressure gave a better distillation temperature and distillation rate than at 8 mBar. It is found that still pot temperature, liquid condensate temperature, vapor temperature, time, and vapor velocity for 4 mBar pressure were better than 8 mBar pressure, since they gave lower above values with a higher vapor velocity as shown in the above mentioned figures. The experimental results for 2 and 3 mBar pressure gave poor results due to sucking of vapor and hence produced erratic data, therefore 4 mBar pressure gave the best lowest pressure with satisfactory results. Pressure above 8 mBar gives higher values of temperature and these results are considered to be un-

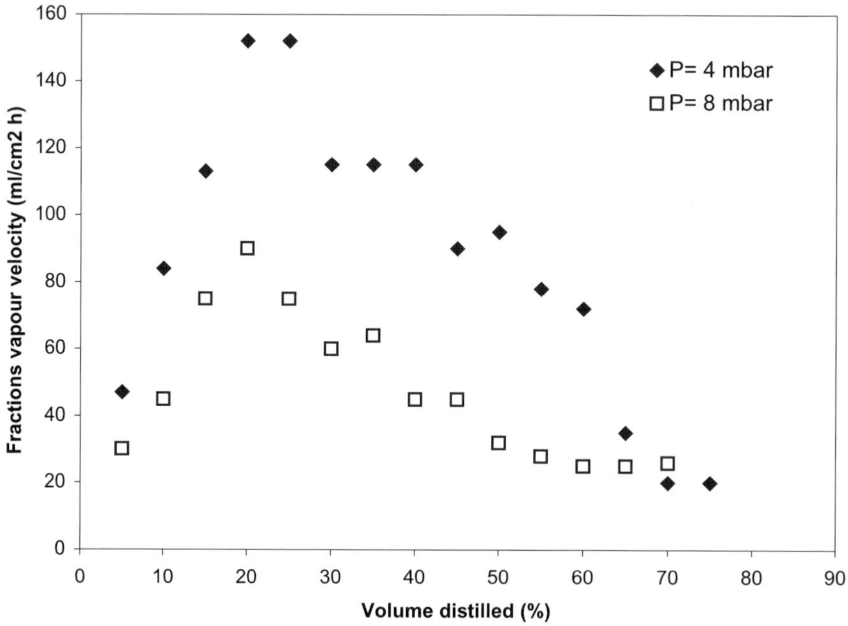

Fig. 4.59. Volume distilled vs. fractions vapor velocity for solvent treated oil at different pressures, $R_r = 2$, heat rate = 600 W, cooling temperature = 15 °C, charge = 700 mL and reflux time base = 0.5 s.

Table 4.20. Tests analysis for fraction distilled for solvent treated oil at vacuum pressure of 8 mBar, reflux ratio of 2, and reflux time base of 0.5 s

Test	Oil base 1	Oil base 2
Sp gr	0.886	0.890
Flash point (°C)	175	198
Pour point (°C)	−15	−9
Viscosity 40 °C (cSt)		39.6
Acid No.	0.22	
Copper corrosion		1b

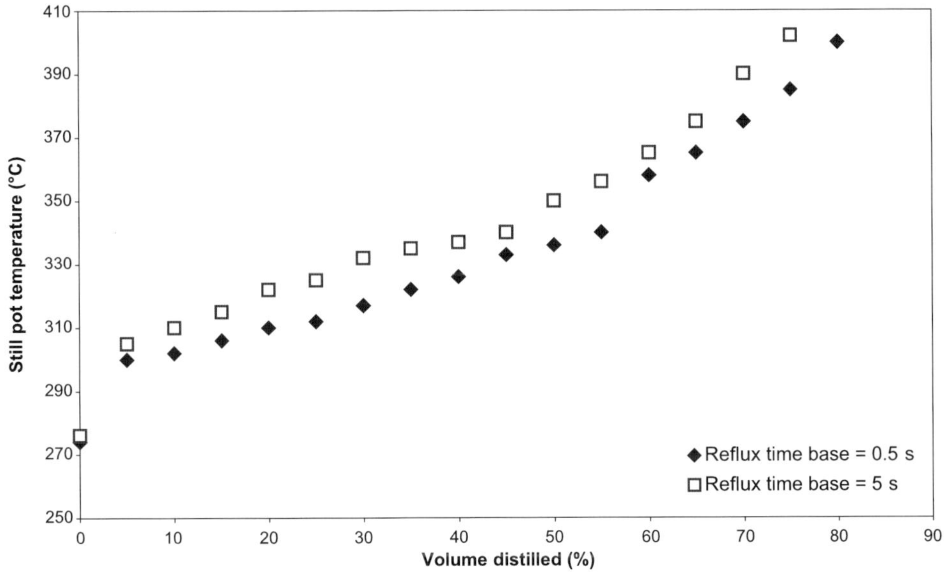

Fig. 4.60. Volume distilled vs. still pot temperature for solvent treated oil at different reflux time base, $P = 4$ mBar, $R_r = 0.5$, heat rate = 600 W, and charge = 700 mL.

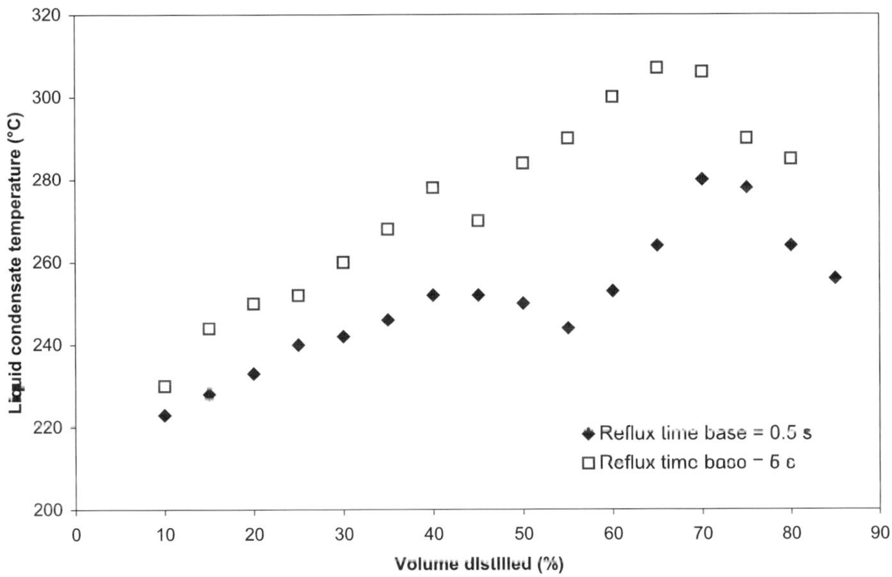

Fig. 4.61. Volume distilled vs. liquid condensate temperature for solvent treated oil at different reflux time base, $P = 4$ mBar, $R_r = 0.5$, heat rate = 600 W, and charge = 700 mL.

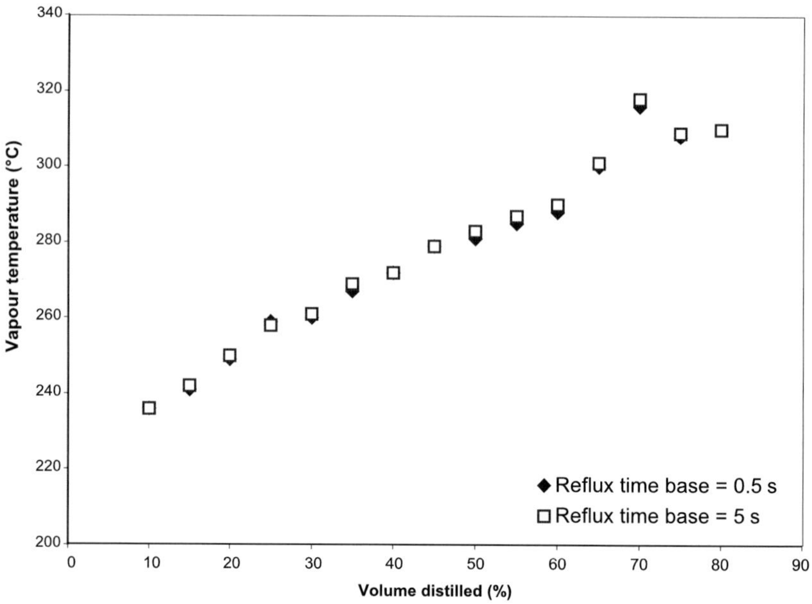

Fig. 4.62. Volume distilled vs. vapor temperature for solvent treated oil at different reflux time base, $P = 4$ mBar, $R_r = 0.5$, heat rate = 600 W, and charge = 700 mL.

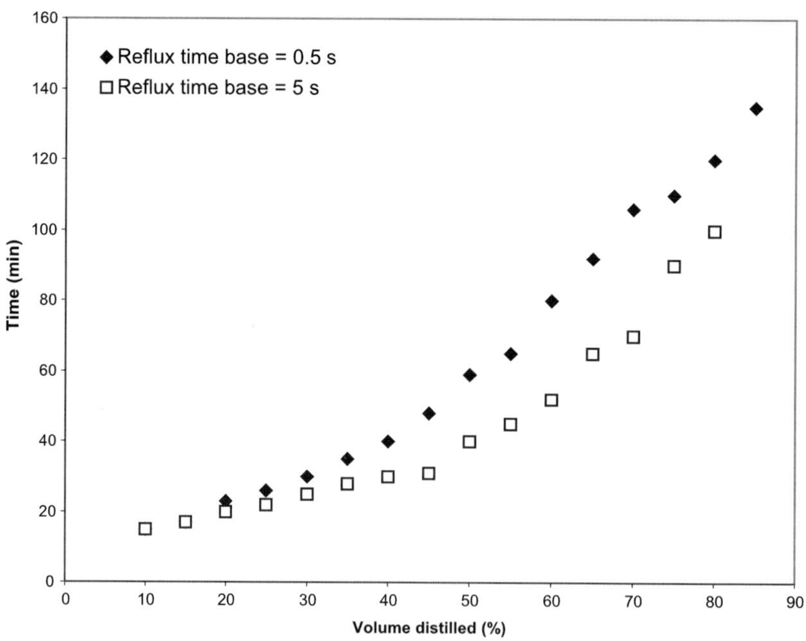

Fig. 4.63. Volume distilled vs. time for solvent treated oil at different reflux time base, $P = 4$ mBar, $R_r = 0.5$, heat rate = 600 W, and charge = 700 mL.

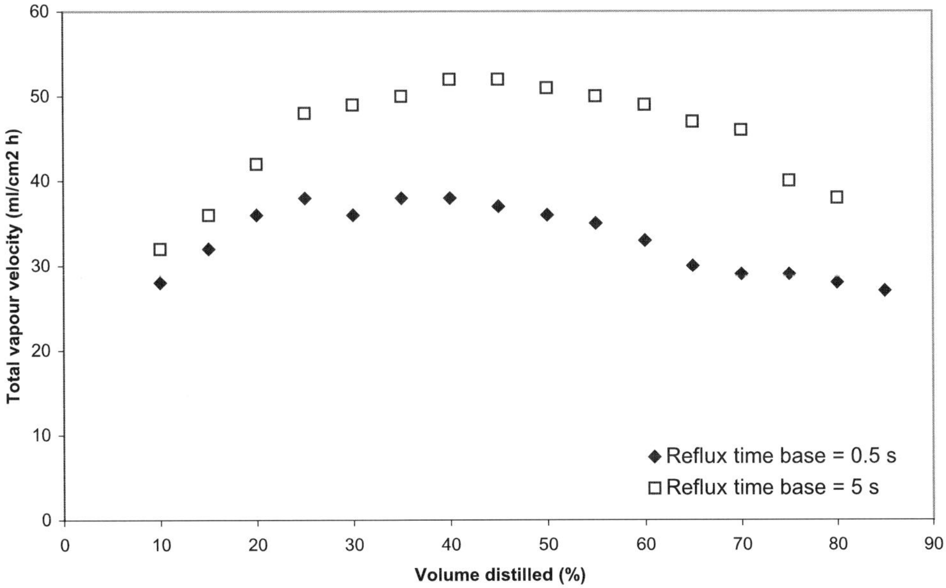

Fig. 4.64. Volume distilled vs. total vapor velocity for solvent treated oil at different reflux time base, $P = 4$ mBar, $R_r = 0.5$, heat rate = 600 W, and charge = 700 mL.

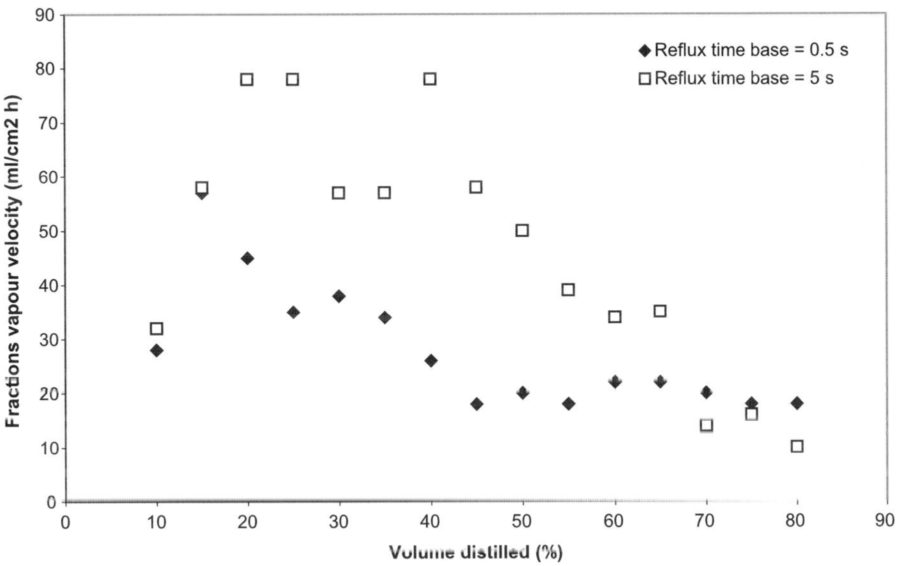

Fig. 4.65. Volume distilled vs. fractions vapor velocity for solvent treated oil at different reflux time base, $P = 4$ mBar, $R_r = 0.5$, heat rate = 600 W, and charge = 700 mL.

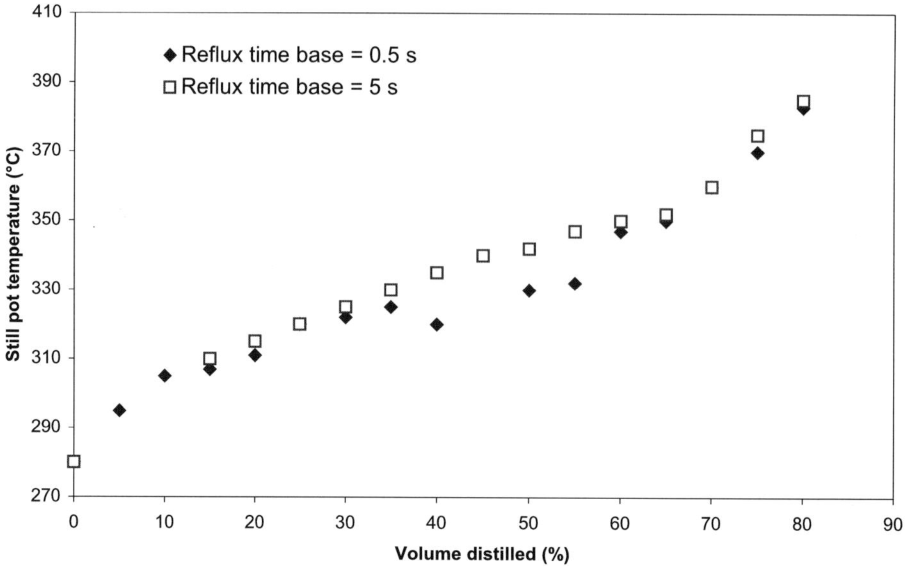

Fig. 4.66. Volume distilled vs. still pot temperature for solvent treated oil at different reflux time base, $P = 4$ mBar, $R_r = 2$, heat rate = 600 W, and charge = 700 mL.

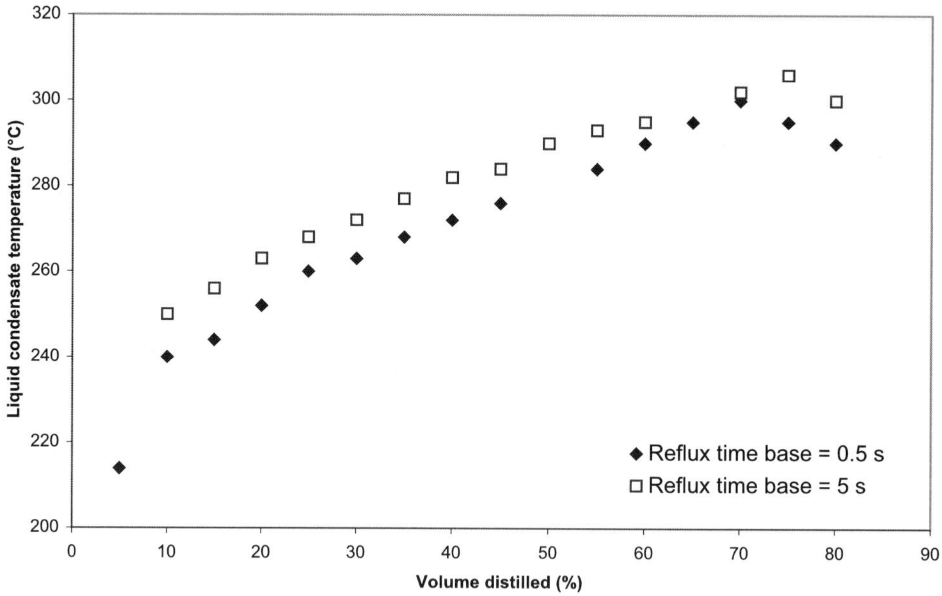

Fig. 4.67. Volume distilled vs. liquid condensate temperature for solvent treated oil at different reflux time base, $P = 4$ mBar, $R_r = 2$, heat rate = 600 W, and charge = 700 mL.

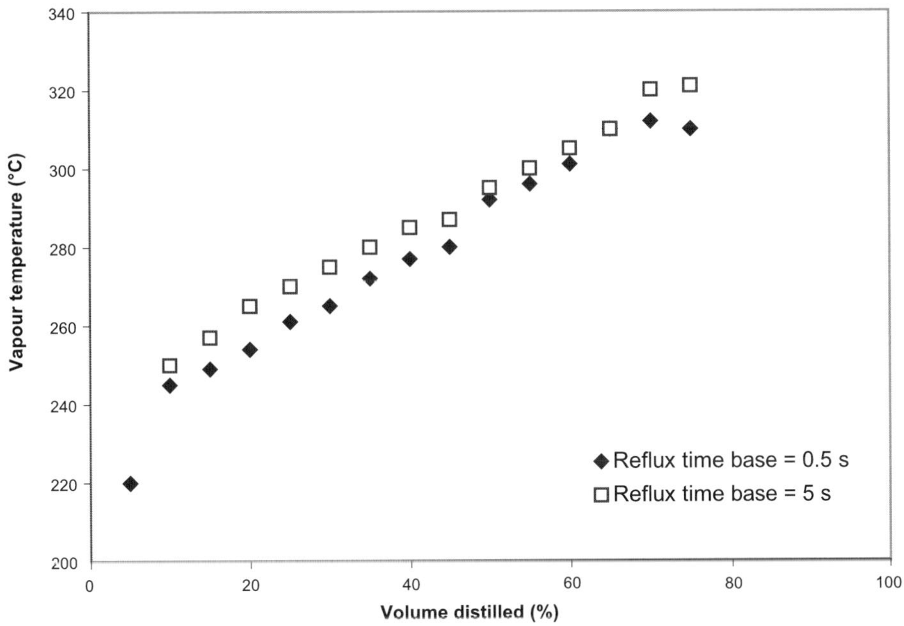

Fig. 4.68. Volume distilled vs. vapor temperature for solvent treated oil at different reflux time base, $P = 4$ mBar, $R_r = 2$, heat rate = 600 W, and charge = 700 mL.

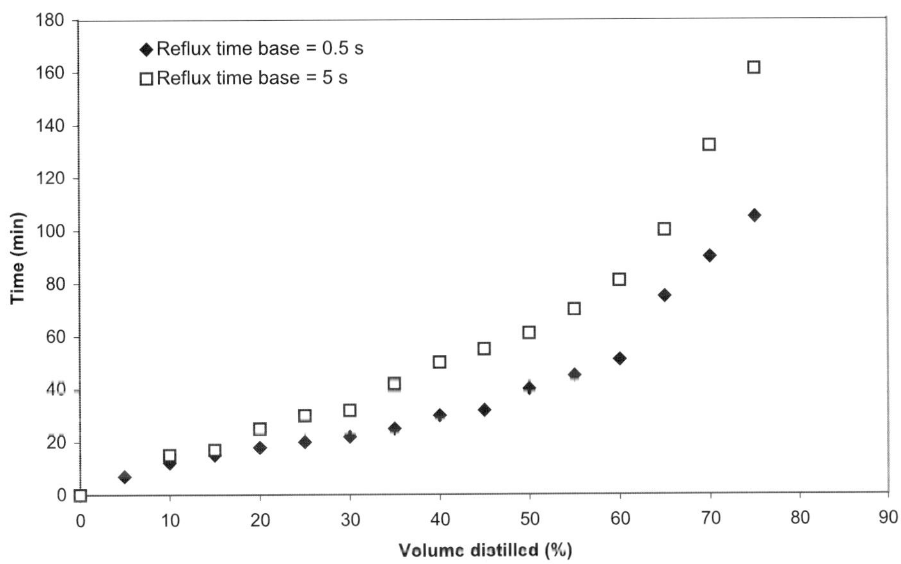

Fig. 4.69. Volume distilled vs. time for solvent treated oil at different reflux time base, $P = 4$ mBar, $R_r = 2$, heat rate = 600 W, and charge = 700 mL.

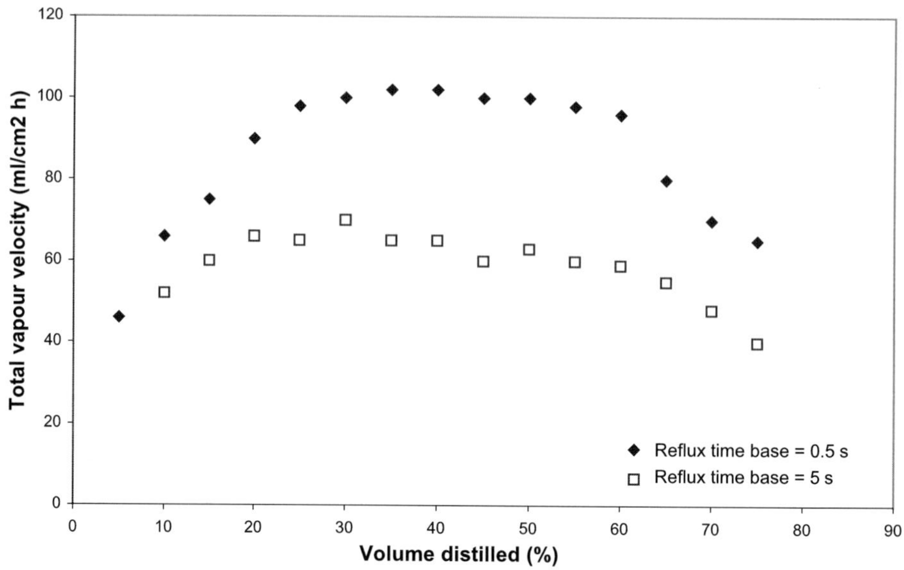

Fig. 4.70. Volume distilled vs. total vapor velocity for solvent treated oil at different reflux time base, $P = 4$ mBar, $R_r = 2$, heat rate = 600 W, and charge = 700 mL.

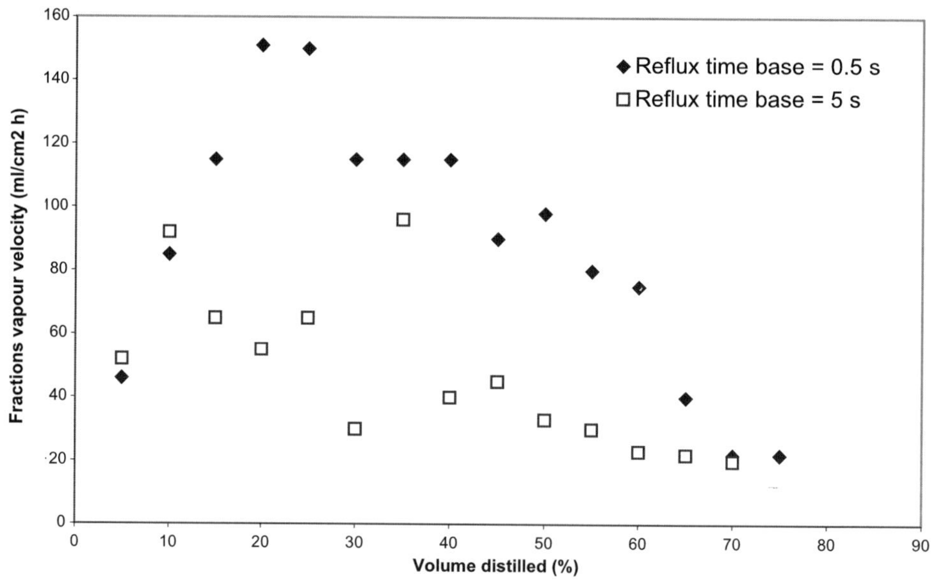

Fig. 4.71. Percent volume distilled vs. fractions vapor velocity for solvent treated oil at different reflux time base, $P = 4$ mBar, $R_r = 2$, heat rate = 600 W, and charge = 700 mL.

satisfactory. At 8 mBar pressure foaming problems appear. The oil foams tend to carry over the liquid oil to the vacuum lines. This problem was eliminated by adding 1% of anti-foam additives. The foaming problem improves when the vacuum pump evacuates the system to recover the pressure drop obtained due to the high rate of evaporation. This evacuation disturbed to the liquid and increased the foaming rate.

4.4.3.3 *Reflux time base.* The fractionation behaviors for solvent treated oil at different reflux time base, i.e. 0.5 s and 5 s, are compared as shown in Figs. 4.60–4.65 for reflux ratio of 2. Similarly Figs. 4.66–4.71 show the fractionation behaviors of solvent treated oil at different reflux time base for reflux ratio of 0.5.

Tables 4.15, 4.18 and 4.19 indicate that the reflux time base does not affect the amount of distillates in the oil. Also these tables show that the two reflux ratios, i.e. 0.5 and 2 at reflux time base of 5 s produce better pour point distillates. A reflux time base of 5 s gave better oil quality than a reflux time base of 0.5 s, while a reflux time base of 0.5 s gave a higher distillation rate than a reflux time base of 5 s, where the reflux ratio for both cases was 2.

4.5 Comparison of the Properties of Used Oil, Virgin Oil, and Used Solvent Treated Oil Distillates

Comparison of the distillates was conducted at two different vacuum pressures. For 4 mBar as shown in Figs. 4.72–4.77 and for 8 mBar as shown in Figs. 4.78–4.83 with all other distillation conditions constant.

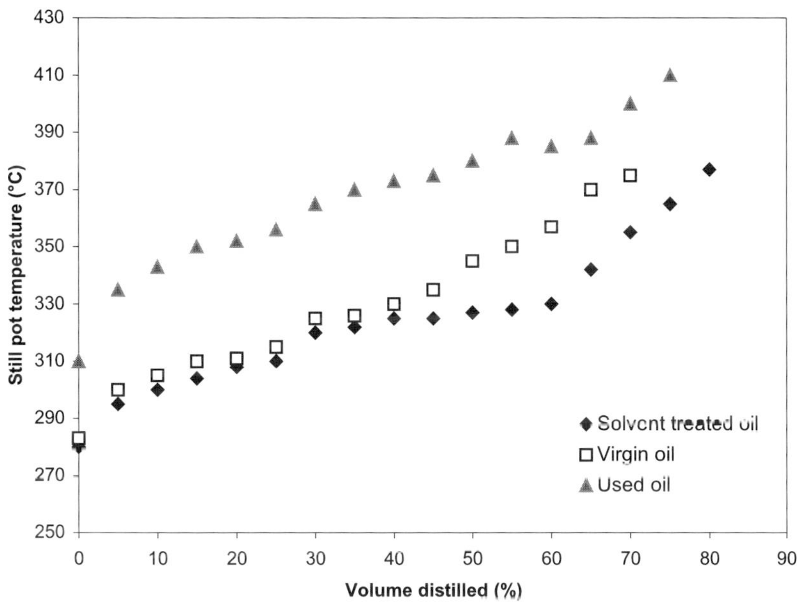

Fig. 4.72. Volume distilled vs. still pot temperature of used, virgin and solvent treated oil, $P = 4$ mBar, $R_r = 2$, heat rate = 600 W, reflux time base = 0.5 s, cooling temperature = 15 °C, and charge = 700 mL.

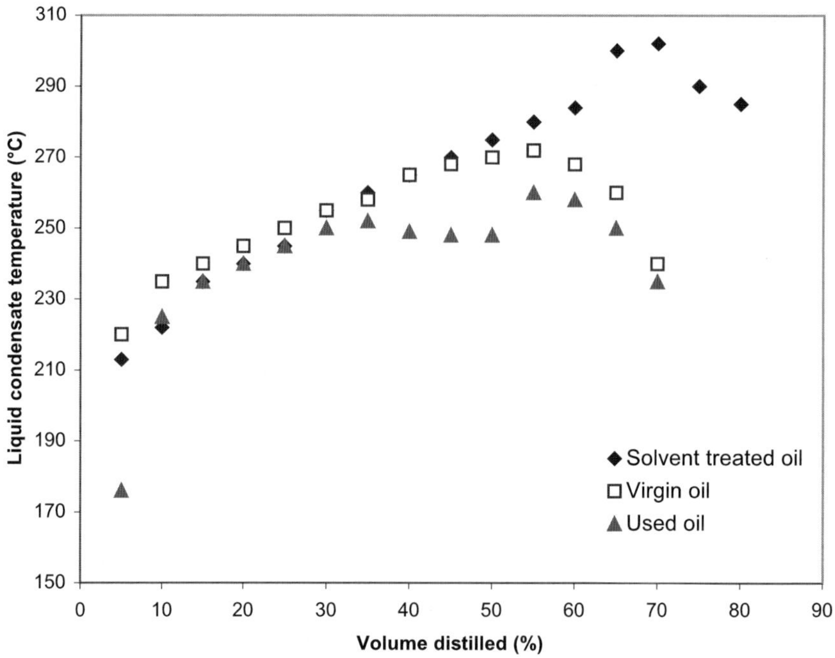

Fig. 4.73. Volume distilled vs. liquid condensate temperature of used, virgin and solvent treated oil, $P = 4$ mBar, $R_r = 2$, heat rate = 600 W, reflux time base = 0.5 s, cooling temperature = 15 °C, and charge = 700 mL.

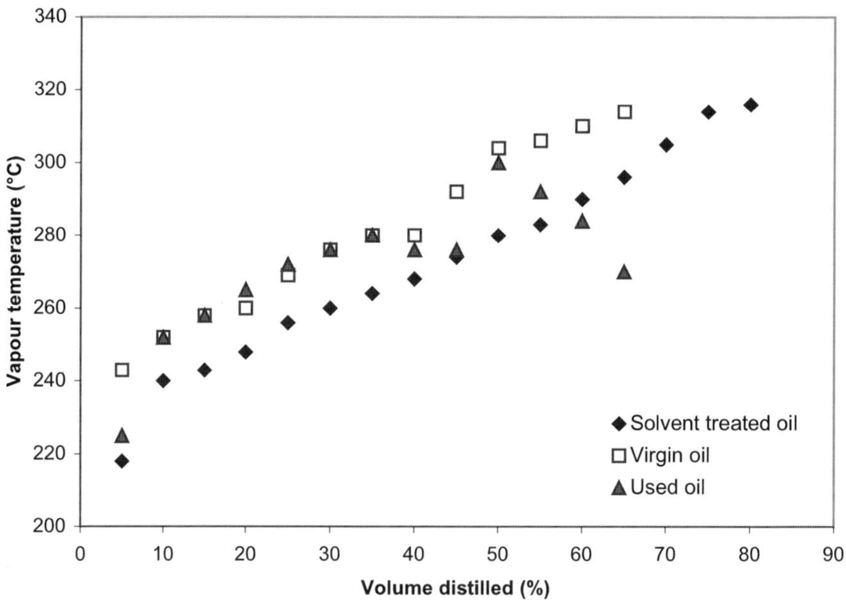

Fig. 4.74. Volume distilled vs. vapor temperature of used, virgin and solvent treated oil, $P = 4$ mBar, $R_r = 2$, heat rate = 600 W, reflux time base = 0.5 s, cooling temperature = 15 °C, and charge = 700 mL.

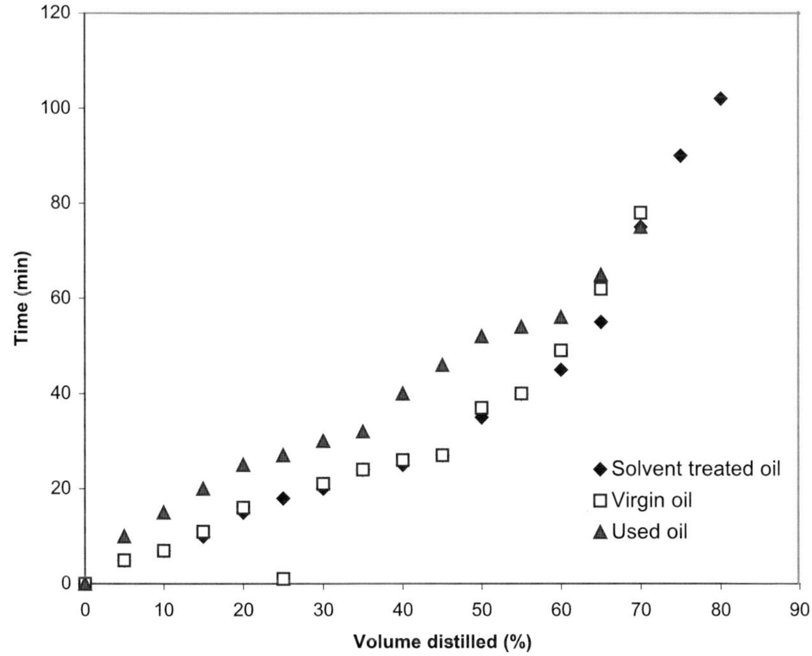

Fig. 4.75. Volume distilled vs. time of used, virgin and solvent treated oil, $P = 4$ mBar, $R_r = 2$, heat rate = 600 W, reflux time base = 0.5 s, cooling temperature = 15 °C, and charge = 700 mL.

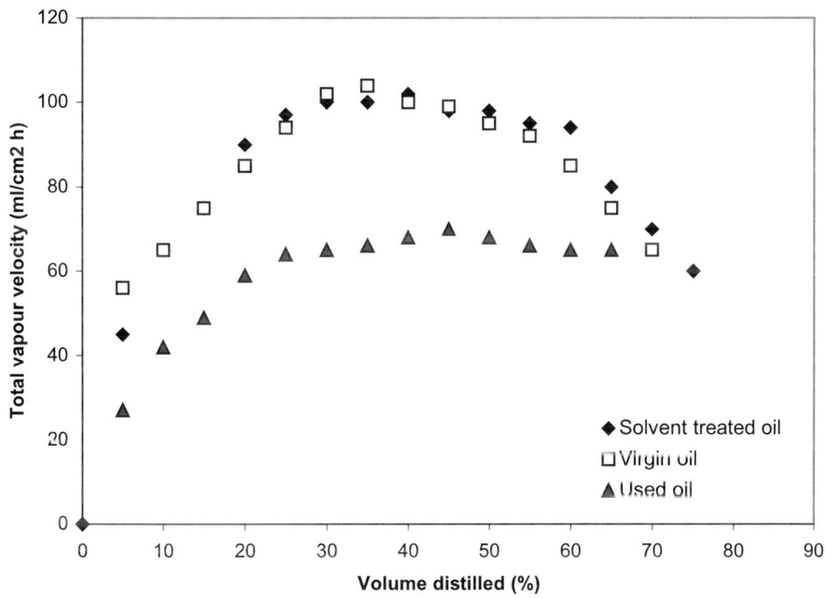

Fig. 4.76. Volume distilled vs. total vapor velocity of used, virgin and solvent treated oil, $P = 4$ mBar, $R_r = 2$, heat rate = 600 W, reflux time base = 0.5 s, cooling temperature = 15 °C, and charge = 700 mL.

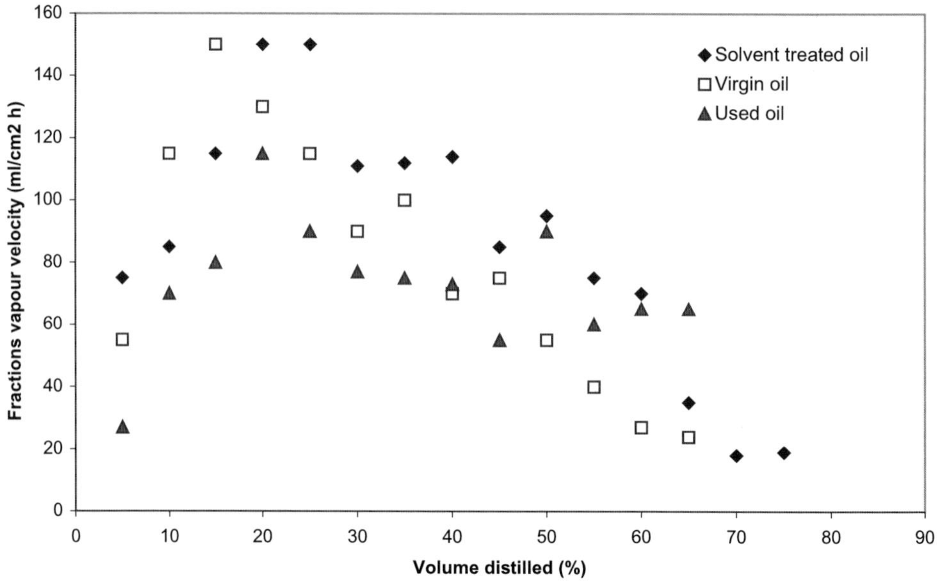

Fig. 4.77. Volume distilled vs. fractions vapor velocity of used, virgin and solvent treated oil, $P = 4$ mBar, $R_r = 2$, heat rate = 600 W, reflux time base = 0.5 s, cooling temperature = 15 °C, and charge = 700 mL.

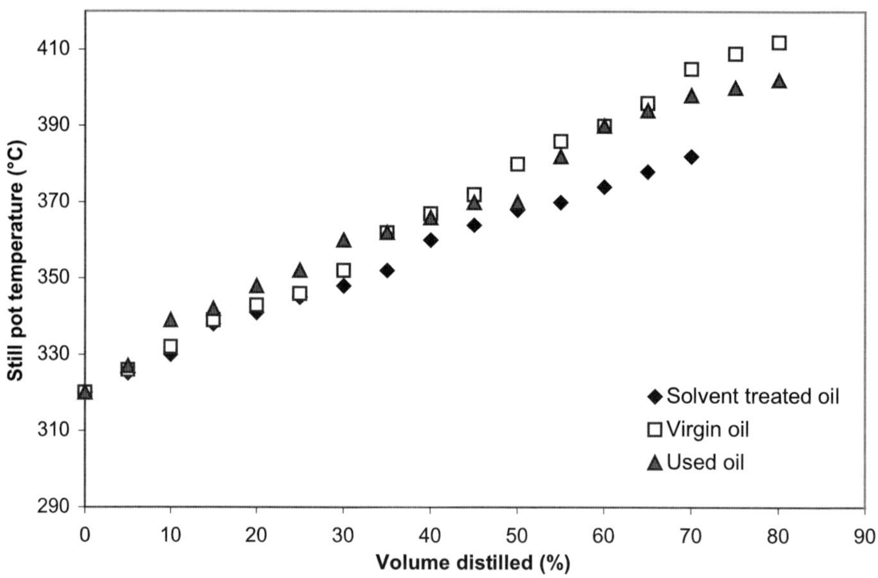

Fig. 4.78. Volume distilled vs. still pot temperature of used, virgin and solvent treated oil, $P = 8$ mBar, $R_r = 2$, heat rate = 600 W, reflux time base = 0.5 s, cooling temperature = 15 °C, and charge = 700 mL.

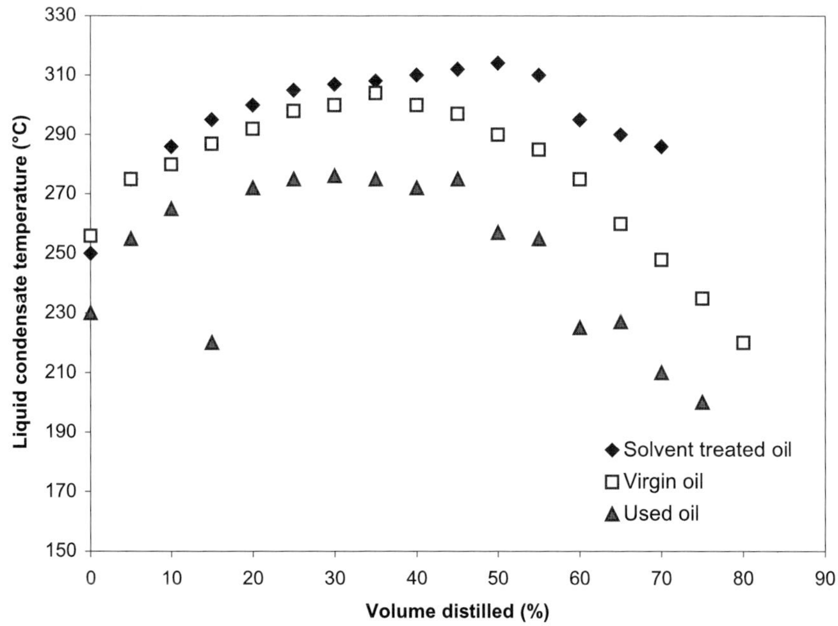

Fig. 4.79. Volume distilled vs. liquid condensate temperature of used, virgin and solvent treated oil, $P = 8$ mBar, $R_r = 2$, heat rate = 600 W, reflux time base = 0.5 s, cooling temperature = 15 °C, and charge = 700 mL.

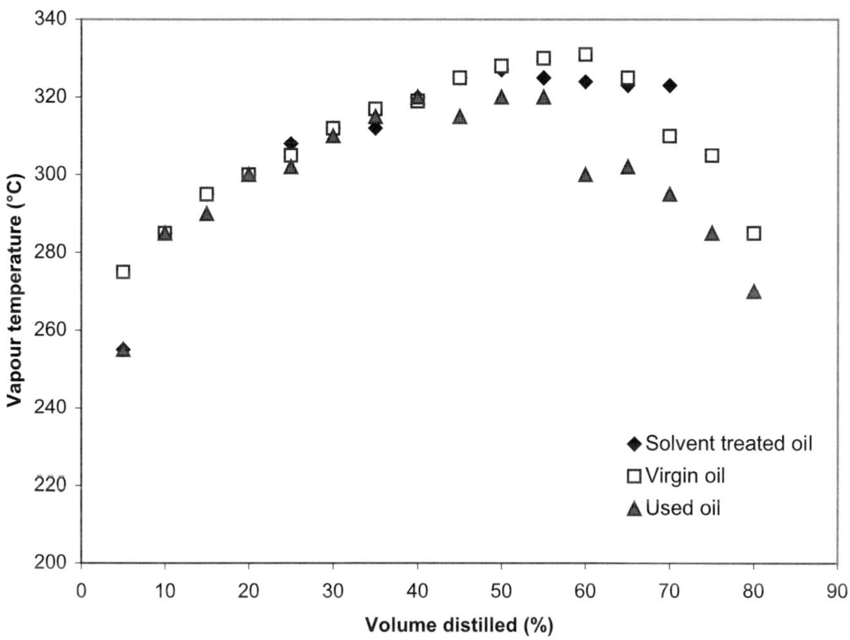

Fig. 4.80. Volume distilled vs. vapor temperature of used, virgin and solvent treated oil, $P = 8$ mBar, $R_r = 2$, heat rate = 600 W, reflux time base = 0.5 s, cooling temperature = 15 °C, and charge = 700 mL.

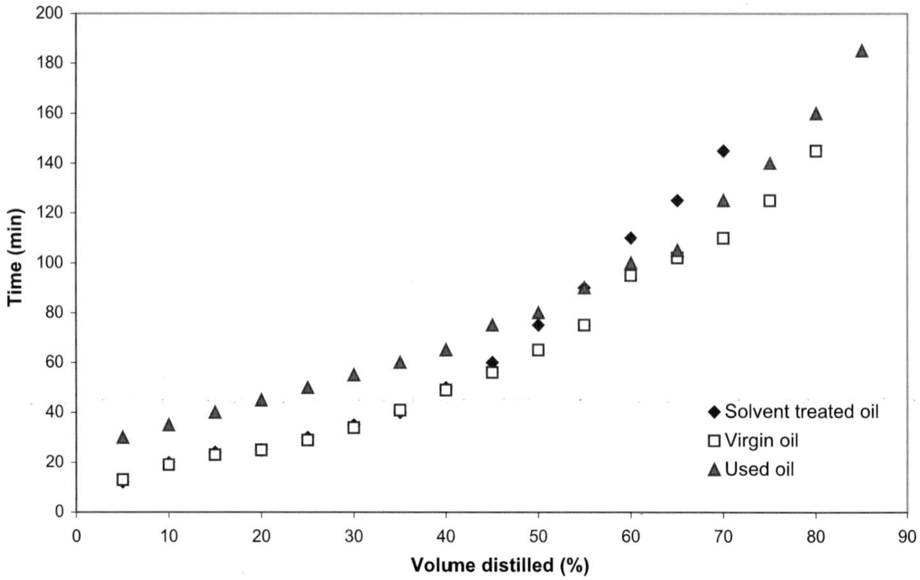

Fig. 4.81. Volume distilled vs. time of used, virgin and solvent treated oil, $P = 8$ mBar, $R_r = 2$, heat rate = 600 W, reflux time base = 0.5 s, cooling temperature = 15 °C, and charge = 700 mL.

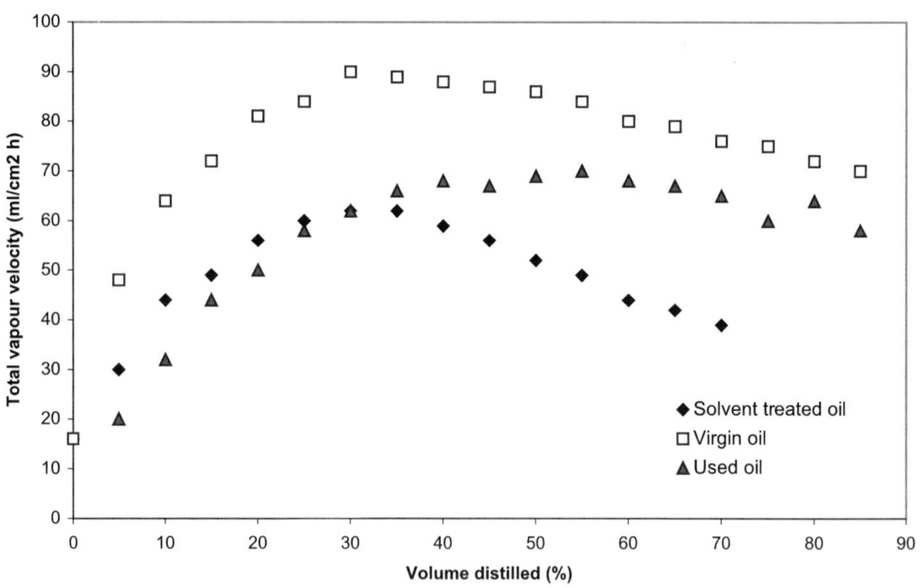

Fig. 4.82. Volume distilled vs. total vapor velocity of used, virgin and solvent treated oil, $P = 8$ mBar, $R_r = 2$, heat rate = 600 W, reflux time base = 0.5 s, cooling temperature = 15 °C, and charge = 700 mL.

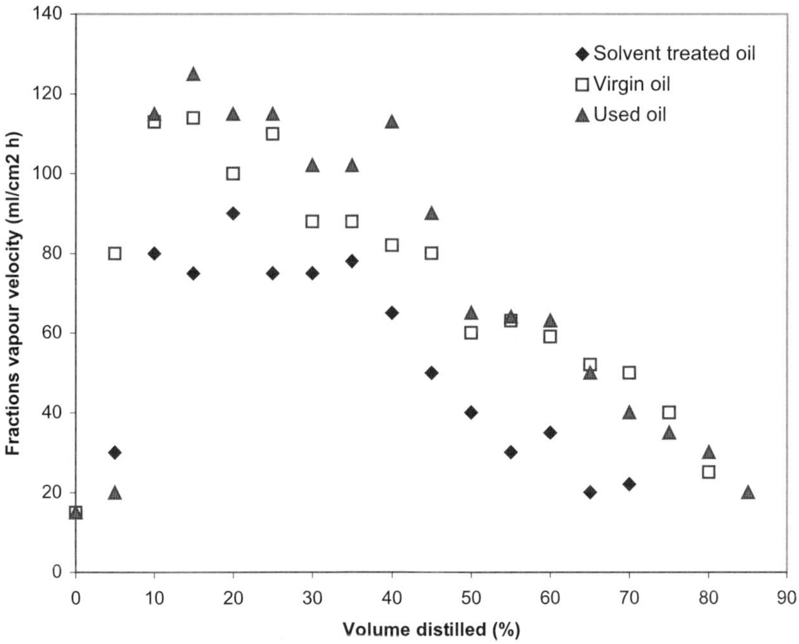

Fig. 4.83. Volume distilled vs. fractions load of used, virgin and solvent treated oil, $P = 8$ mBar, $R_r = 2$, heat rate = 600 W, reflux time base = 0.5 s, cooling temperature = 15 °C, and charge = 700 mL.

For the two vacuum pressures, all the distillate oil properties of the solvent treated oil are better than the distillates of the used oil as indicated by Tables 4.9, 4.15, and 4.16. While Tables 4.8 and 4.14 indicate that the used oil and solvent treated oil produce nearly similar distillate amounts. The fractionation behaviors of the three oils as shown in Figs. 4.72–4.77 for 4 mBar pressure indicate that solvent treated oil gave a better distillation temperature and distillation rate than other oils. As indicated in Figs. 4.78–4.83 for 8 mBar pressure virgin oil gave better distillation temperature and distillation rate fraction in solvent treated oil and in used oil.

No comparison was reported for the residues between the three oils. At a vacuum pressure of 4 mBar, distillate properties and distillation behavior were nearly identical between solvent treated oil and virgin oil, but different than used oil. At 8 mBar pressure operation, the solvent treated oil has almost same distillate properties, but lower distillation rate than at a vacuum pressure of 4 mBar. While virgin oil at 8 mBar pressure gave a lower rate of distillation than that at 4 mBar pressure. However, virgin oil gave a better distillation rate than solvent treated oil at 8 mBar pressure. Used oil gave better distillation results at 8 mBar than 4 mBar, since there were no losses of the vapor due to sucking. It is well known that used oil gave unsatisfactory results for both vacuum pressures with respect to solvent treated oil and virgin oil.

The still pot temperature for solvent treated oil distillates is lower than that of virgin oil and used oil distillates at 4 mBar pressure. The still pot temperature of solvent treated oil is similar to that of virgin oil, but differs from still pot temperature of used oil. Solvent treated

Table 4.21. Tests analysis of oil distillates properties for used oil, virgin oil and solvent treated oil

Test	Used oil distillates		Virgin oil distillates	Solvent treated oil distillates	
	Oil base 1	Oil base 2	Oil base 1	Oil base 1	Oil base 2
Sp gr	0.903	0.914	0.885	0.888	0.897
Viscosity 40 °C (cSt)	39.3	36.1	–	37.0	41.2
Viscosity 100 °C (cSt)	6.4	5.7	–	–	9.5
Flash point (°C)	195	180	–	200	200
Pour point (°C)	−5	1	−1	−7	−2
Ash (wt%)	0.26	0.13	0.13	0	0.02
Sulfur (wt%)	1.14	1.66	–	–	–
Acid No.	–	–	–	0.04	0.26
Water (wt%)	0	0	0	0	0
Saponification No.	–	–	5	–	–
Copper corrosion	–	–	–	1a	1a
Viscosity index	113	95	–	–	161

Table 4.22. Metal contents of used oil, residue, virgin oil, and solvent treated oil

Metals	Used oil (ppm)	Residue (ppm)	Virgin oil (ppm)	Solvent treated oil (ppm)
Na	15.8	17.85	60.4	8.3
Zn	2282	1482.14	35.2	19.60
Ni	31.3	42.5	0.00	1.2
Mg	13.18	3.7	101.3	3.10
Mn	5.5	3.9	1	0.2
Cr	0.00	0.00	0.00	0.00
Cu	8.6		3.47	1.1
Ca	168.3		208.5	0.1
Fe	711.6	326.8	5.6	17.3
K	4.7		78.5	6.1

oil gave better results than virgin oil and used oil due to removal of heavy components and additives as shown in Fig. 4.72. Liquid condensate temperature and vapor temperature for the solvent treated oil and virgin oil gave similar behaviors and both differ from that of used oil as shown in Figs. 4.73 and 4.74. The rate of distillation of solvent treated oil and virgin oil up to 60% was similar. Above 60% the solvent treated oil gave higher distillation rate than virgin oil due to the removal of additives that have high boiling point and high molecular weights, from the solvent treated oil by previous treatment with solvents. Both virgin oil and solvent treated oil gave better distillation rate and vapor velocity than used oil as shown in Figs. 4.75–4.77.

At 8 mBar pressure, the results are not representative of the actual behavior, since an 8 mBar vacuum pressure was not enough to fractionate clearly the solvent treated oil as indicated by the results in the figures and only a test analysis can be used to give definite behavior of quality of best oil cuts. It was found that liquid condensate temperature, vapor temperature and fraction vapor velocity did not represent generally the behavior of oil cuts as shown in Figs. 4.79, 4.80 and 4.83, while Fig. 4.78 indicates that still pot temperature for virgin oil is better than still pot temperature of used oil. Figure 4.81 indicates that time distilled for virgin oil is better than time distilled for used oil, while Fig. 4.82 shows that total vapor velocity of virgin oil is better than that of used oil.

4.6 Comparison between Used Oil, Virgin Oil and Re-refined Oil

A comparison between used oil, virgin oil and re-refined oil, i.e. solvent treated oil formulated with additives, was carried out as shown in Table 4.23. Table 4.21 shows that the

Table 4.23. Tests analysis for used oil, virgin oil, and re-refined oil formulated with additives

Test	Used oil	Virgin oil	Re-refined oil	
			5% additives	10% additives
Sp gr	0.905	0.890	0.890	0.890
Viscosity (cSt) 40 °C	71.1	101.0	54.2	75.6
Viscosity (cSt) 100 °C	9.0	11.5	8.0	9.9
Flash point (°C)	155	220	210	210
Pour point (°C)	−6	−8	−18	
Ash (wt%)	0.352	0.13		
Sulfur (wt%)	1.67			
Acid No.	2.55		0.1	0.2
Water (wt%)	2		0	0
Viscosity index	100	100	113	110

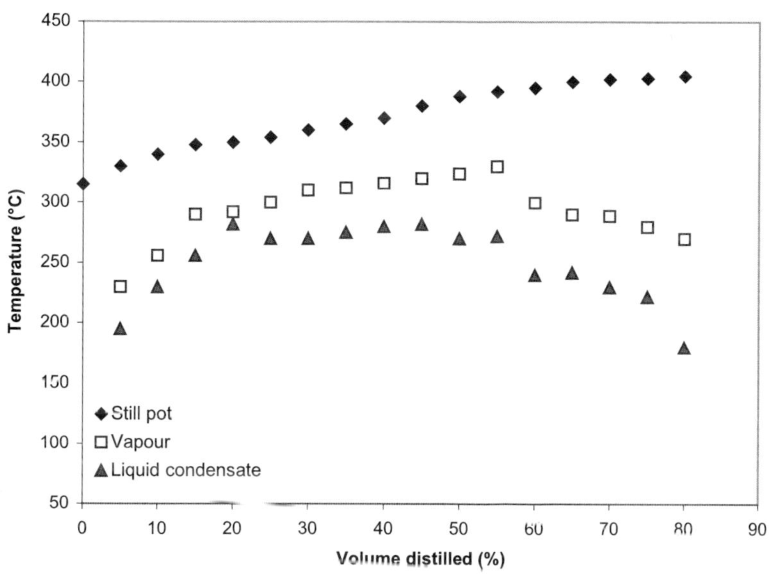

Fig. 4.84. Volume distilled vs. temperature for used oil at 8 mBar pressure, $R_r = 2$, reflux time base = 0.5 s, heat rate = 600 W, cooling temperature = 15 °C.

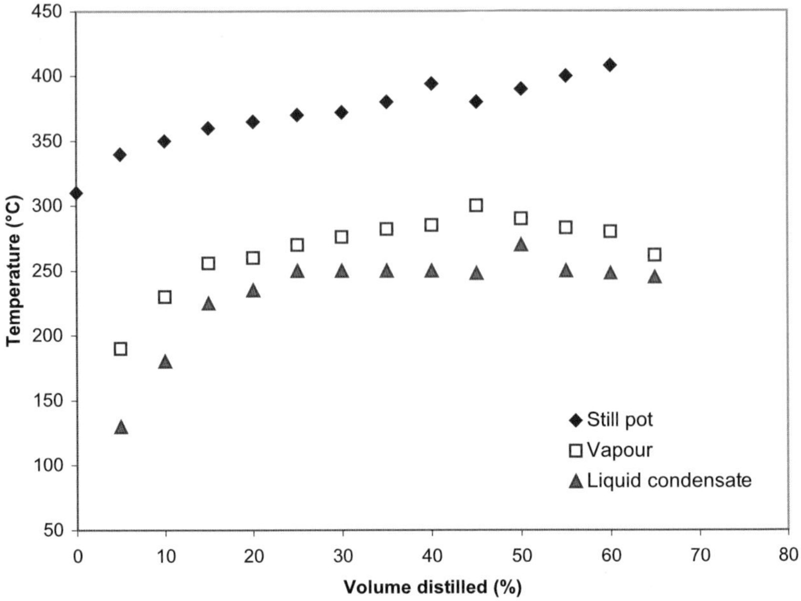

Fig. 4.85. Volume distilled vs. temperature for used oil at 4 mBar pressure, $R_r = 2$, reflux time base = 0.5 s, heat rate = 600 W, cooling temperature = 15 °C.

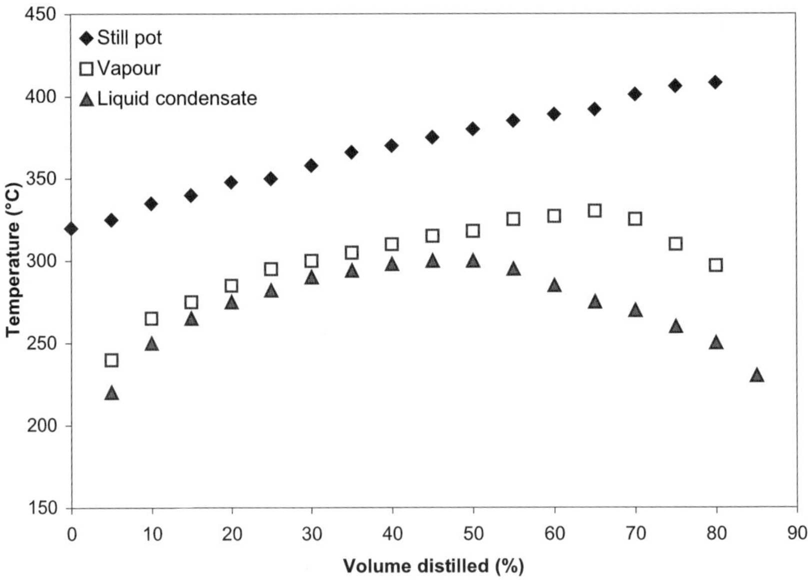

Fig. 4.86. Volume distilled vs. temperature for virgin oil at 8 mBar pressure, $R_r = 2$, reflux time base = 0.5 s, heat rate = 600 W, cooling temperature = 15 °C.

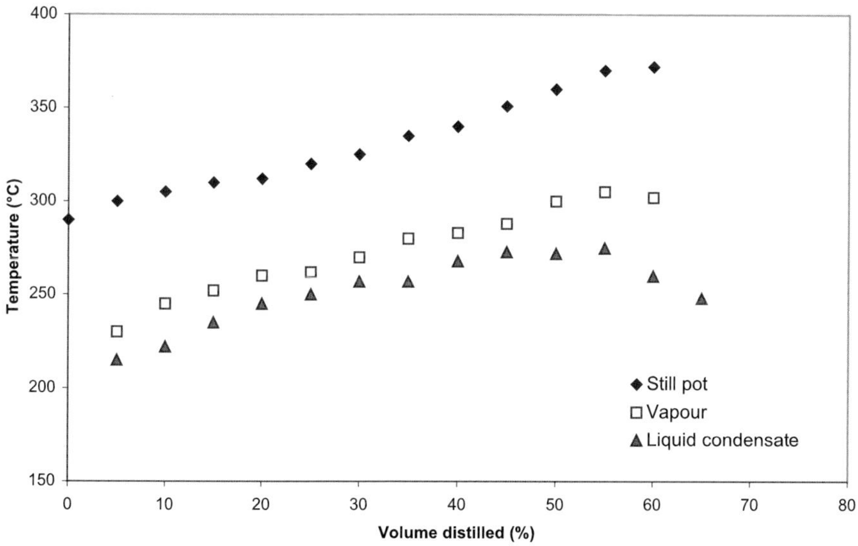

Fig. 4.87. Volume distilled vs. temperature for virgin oil at 4 mBar pressure, $R_r = 2$, reflux time base = 0.5 s, heat rate = 600 W, cooling temperature = 15 °C.

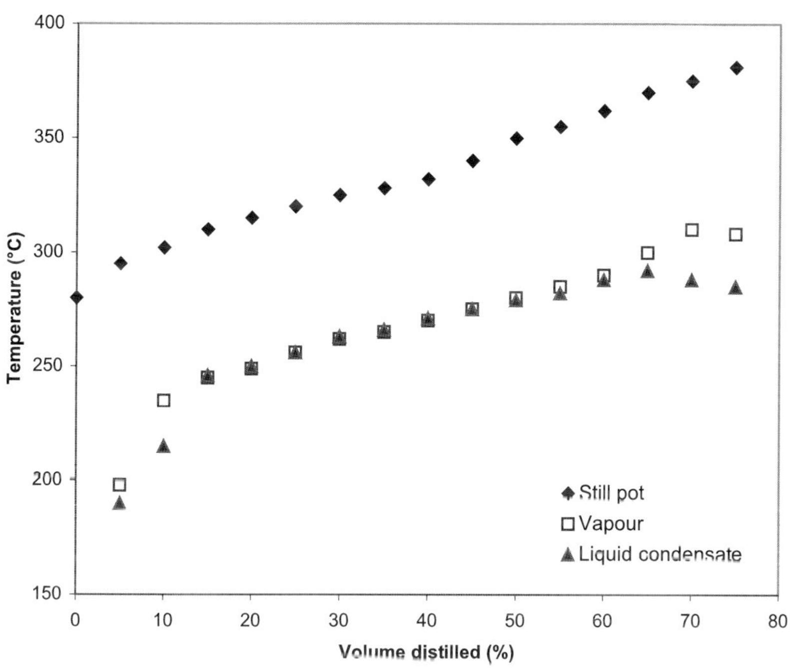

Fig. 4.88. Volume distilled vs. temperature for solvent treated oil at 4 mBar pressure, $R_r = 2$, reflux time base = 0.5 s, heat rate = 600 W, cooling temperature = 15 °C.

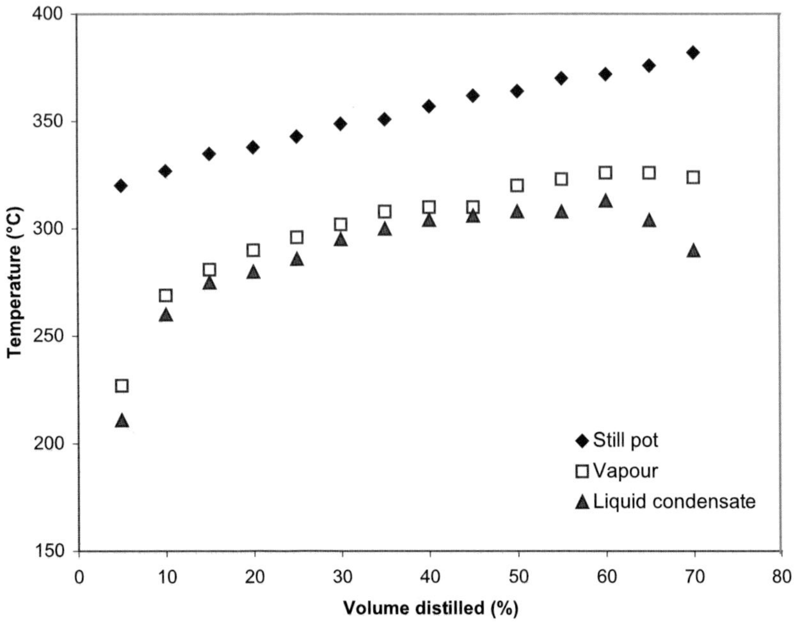

Fig. 4.89. Percent volume distilled vs. temperature for solvent treated oil at 8 mBar pressure, $R_r = 2$, reflux time base = 0.5 s, heat rate = 600 W, cooling temperature = 15 °C.

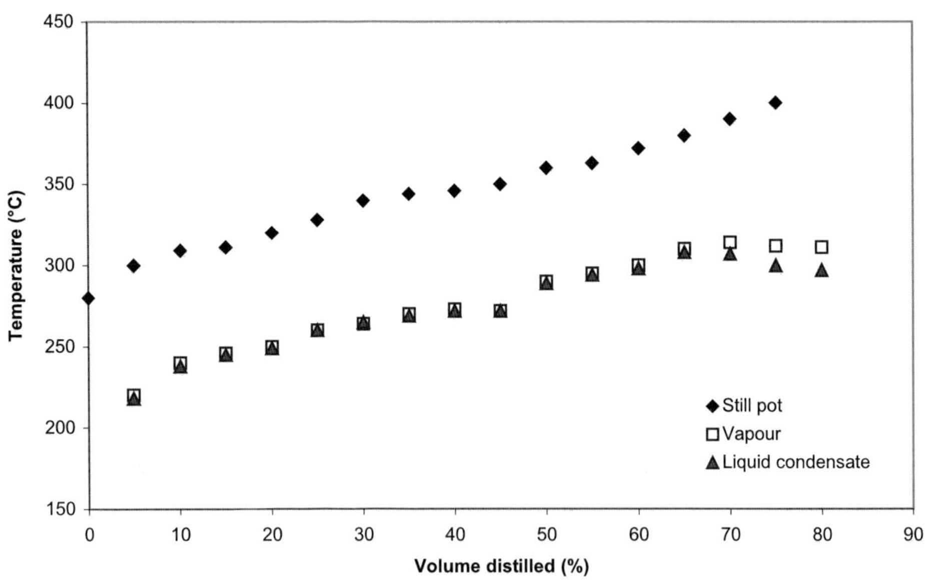

Fig. 4.90. Volume distilled vs. temperature for solvent treated oil at 4 mBar pressure, $R_r = 0.5$, reflux time base = 5 s, heat rate = 600 W, cooling temperature = 15 °C.

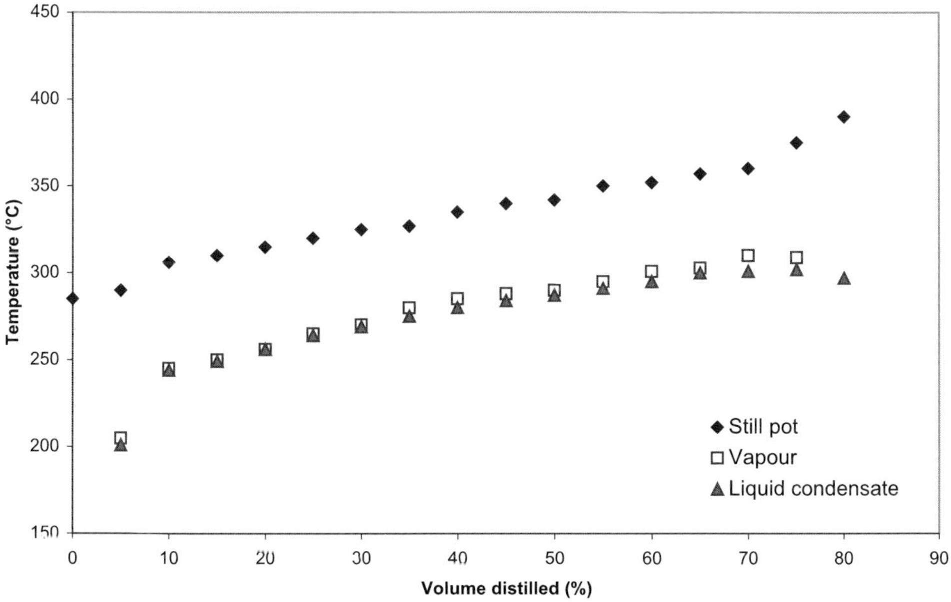

Fig. 4.91. Volume distilled vs. temperature for solvent treated oil at 4 mBar pressure, $R_r = 2$, reflux time base = 5 s, heat rate = 600 W, cooling temperature = 15 °C.

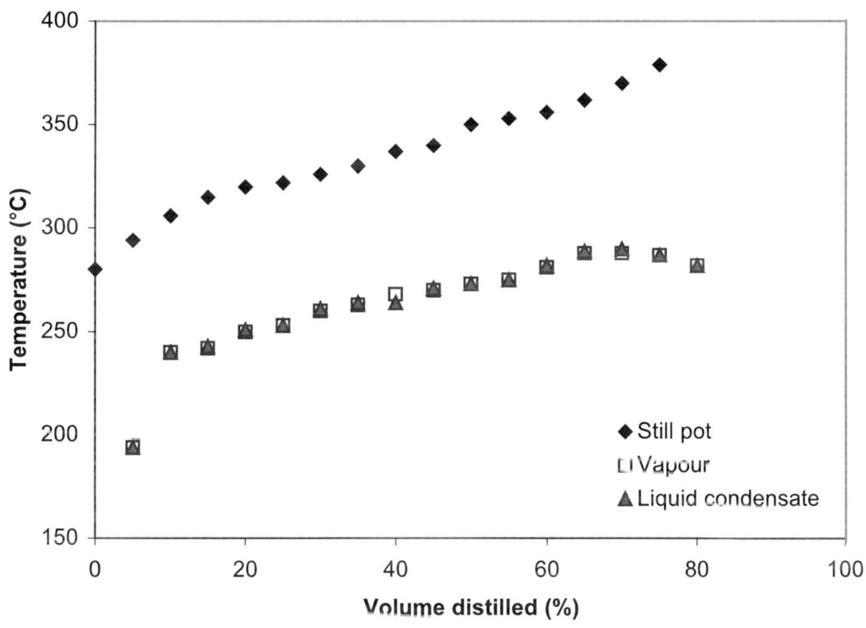

Fig. 4.92. Volume distilled vs. temperature for solvent treated oil at 4 mBar pressure, $R_r = 3$, reflux time base = 5 s, heat rate = 600 W, cooling temperature = 15 °C.

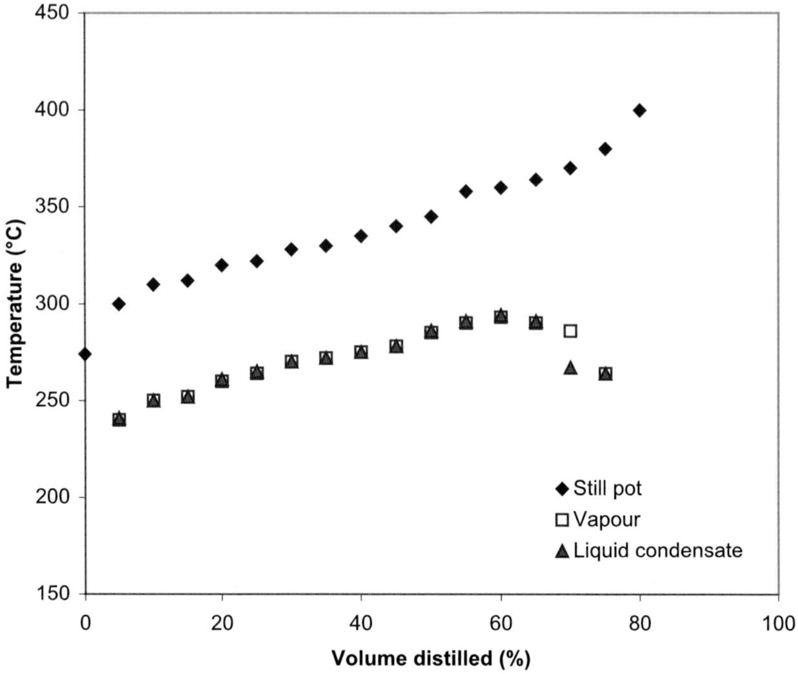

Fig. 4.93. Volume distilled vs. temperature for solvent treated oil at 4 mBar pressure, $R_r = 4$, reflux time base = 5 s, heat rate = 600 W, cooling temperature = 15 °C.

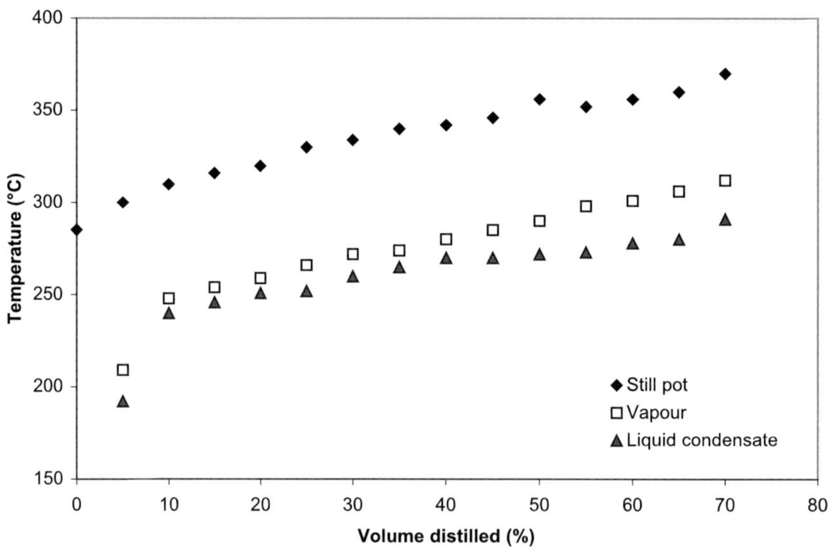

Fig. 4.94. Volume distilled vs. temperature for solvent treated oil at 4 mBar pressure, $R_r = 6$, reflux time base = 5 s, heat rate = 600 W, cooling temperature = 15 °C.

solvent treated oil without additives is more pure than other oils as indicated by very low acid number and ash content. So this base oil is suitable for the additive formulation to be re-used again with no further process. The additives of 5% of viscosity improver in the solvent treated oil improve, in addition to the viscosity, the pour point to better values.

Table 4.22 shows a comparison of metal content of used oil, asphalted residue of distillation of solvent treated oil, virgin oil and solvent treated oil. The solvent treated oil has very little metal content, which indicates that it is very pure oil. The Zn and Fe element levels in used oil were higher than virgin oil, which indicates that these elements were introduced by the engine. The additives introduced the Ca element. The major elements, i.e. Zn, Fe were reduced by the dehydration, solvent extraction, solvent stripping and vacuum distillation processes to very low values, which indicates the efficiency of the process. The other elements, which are reported in Table 4.22 were traces and they were reduced to minimum values by this process.

4.7 Comparison between Still Pot, Liquid Condensate, and Vapor Temperatures

Special data, i.e. two distillation experiments for used oil, two distillation experiments for virgin oil and seven distillation experiments for solvent treated oil were compared in order to correlate between still pot, liquid condensate and vapor temperatures. Liquid condensate and vapor temperatures are lower than still pot temperatures due to a temperature gradient in the column. Liquid condensate and vapor temperatures are not recorded at the same time for the same fraction. The difference between vapor temperature and liquid condensate temperature for used oil and virgin oil are noticeable as shown in Figs. 4.84–4.87, while there is no clear differences for solvent treated oil as shown in Figs. 4.88–4.94. Liquid condensate temperature is lower than vapor temperature due to subcooling of liquid condensate below condensation temperature. The heavy components in the used oil and virgin oil, which were carried over by vapor to the condenser by noticeable quantity (approximately 6% of distilled oil) were sticky, polymeric in nature, had high molecular weight, and were viscous. These heavy components were removed before distillation in solvent extraction process to produce oil free of these materials therefore the vapor temperature and liquid condensate temperature for solvent treated oil is nearly the same.

Chapter 5

Conclusions

Four process stages were studied, namely: dehydration, solvent extraction, solvent stripping, and vacuum distillation. The study was carried out on a used oil mixture drained from different automobiles. All gasoline and water fractions were separated using vacuum distillation up to 4 mBar at 200 °C for the dehydration process.

Extraction reduces the contaminants to low level, i.e. 65% ash reduction, such that no further operational problems were encountered on vacuum distillation. The best oil recovery and ash reduction by extraction that were obtained using optimum evaluated solvent to oil ratio of 3 to 1 with solvent composition of 40% butanone, 25% 2-propanol, and 35% 1-butanol were 65% ash reduction and 96% oil recovery. These results are slightly different from the mathematical optimization method that gives 69% ash reduction and 98% oil recovery.

Solvent stripping was conducted by two steps namely, atmospheric distillation with temperature as high as 200 °C and vacuum distillation with vacuum pressure down to 40 mBar. Foaming during solvent stripping was reduced to minimum during distillation by controlling operational parameters such as vacuum pressure (120 mBar) by using only vacuum distillation, heat rate (380 W), and charge (1500 mL). About 95% of solvent can be recovered in each complete operation, therefore during a new second run, 5% of make up solvent was required.

Vacuum distillation was carried out to fractionate used oil, virgin oil, and solvent treated oil. It was found that a 600 W heat rate and 700 mL of charge were most suitable for distillation apparatus.

During the dehydration process, the best operating pressure and temperature were 4 mBar and 200 °C respectively. Solvent to oil ratio of 3 to 1 with solvent composition of 40% butanone, 25% 2-propanol and 35% 1-butanol were found to be the optimum composition for solvent extraction. Solvent stripping was conducted by two stages; namely atmospheric distillation to recover butanone and 2-propanol solvents and vacuum distillation at 40 mBar to remove 1-butanol. Controlling the heat rate at 380 W for 1500 mL charge minimized the foaming problem. Cracking reactions do not influence oil properties up to 400 °C during operational times up to 30 minutes.

A reflux ratio of 6 to 1 with reflux time base of 5 s gave better base oil properties with solvent treated oil than other studied reflux ratios, but due to low distillation rate, a reflux ratio of 4 to 1 was considered to be the optimum.

Distillation of solvent treated oil shows that there are no major differences in oil properties when vacuum pressure of 4 mBar and 8 mBar were used, but the rate of distillation and temperature for 4 mBar of pressure were better than for 8 mBar of pressure. Solvent treated oil for 4 mBar of pressure gave a better distillation rate and temperature than virgin oil and used oil, while at 8 mBar virgin oil gave better results than for other oils.

The metal content of used oil was greatly reduced, i.e., Fe from 712 ppm to 17 ppm, and Zn from 2282 ppm to 20 ppm using dehydration, solvent extraction, solvent stripping, and vacuum distillation processes. The best re-refined lubricating oil was obtained by dehydration, solvent extraction, solvent stripping and vacuum distillation processes with specified amount of additives such as 5% and 10% of polymethacrylate, used as viscosity improver and pour point depressant.

All distillates obtained from used oil by fractional distillation were not suitable for re-use.

Appendix A

Distribution Coefficient

Distribution coefficient (K_S) of solvent extraction process is based on the assumption that the ash content of a sample represents the contaminant content:

$$K_S = \frac{\text{contaminants, wt. ratio in extract}}{\text{contaminants, wt. ratio in raffinate}}.$$

Table A.1. For optimum solvent to oil ratio

Solvent to oil ratio	K_S	$1/K_S$
2:1	0.482	2.10
3:1	0.116	8.60
4:1	0.046	21.70
5:1	0.026	38.46
6:1	0.101	9.90

Table A.2. For optimum solvent composition experiment

Exp. No.	K_S	$1/K_S$
1	0.048	20.8
2	0.078	12.8
3	0.083	12.1
4	0.056	17.9
5	0.087	11.5
6	0.050	20.0
7	0.064	15.6
8	0.111	9.0
9	0.135	7.4
10	0.043	23.3
11	0.039	25.6
12	0.044	22.7
13	0.204	5.0
14	0.190	5.3
15	0.082	12.2

Appendix B

Optimization Program

The optimization program is written in Turbo-Basic programming language. The subroutine "2000" represents the penalty objective functions which vary with different bottle's properties.

```
10 PRINT" Optimisation "
12 r=1
15 PRINT"R=":INPUT R(r)
17 PRINT"C=":INPUT C
20 REM Z=F(X1,X2) AT 2000
30 PRINT"NUMBER OF VARIABLES":INPUT N
40 DIM X(N),B(N),Y(N),P(N)
50 PRINT"INITIAL POINT X1,X2,...XN"
60 FOR I=1 TO N:INPUT X(I):NEXT I
70 PRINT"STEP LENGTH":INPUT H
80 K=H
90 FOR I=1 TO N
100 Y(I)=X(I):P(I)=X(I):B(I)=X(I):NEXT I
110 GOSUB 2000:FI=Z
120 PRINT "INITIAL VALUE "Z
130 FOR I=1 TO N:PRINT X(I),:NEXT I:PRINT""
140 PS=0:BS=1
150 REM EXPLORE ABOUT BASE POINT
180 J=1:FB=FI
200 X(J)=Y(J)+K
210 GOSUB 2000
220 IF Z<FI THEN GOTO 280
230 X(J)=Y(J)-K
240 GOSUB 2000
250 IF Z<FI THEN GOTO 280
260 X(J)=Y(J)
270 GOTO 290
280 Y(J)=X(J)
290 GOSUB 2000
300 FI=Z
310 PRINT"EXPLORATION STEP"Z,Q,QQ
```

```
320 FOR I=1 TO N:PRINT X(I),:NEXT I:PRINT""
330 IF J=N THEN GOTO 360
340 J=J+1
350 GOTO 200
360 IF FI<FB-1E-08 THEN GOTO 540
370 REM AFTER 360 MAKE A PATTERN MOVE IF FUNCTION HAS BEEN
    REDUCED
380 IF PS=1 AND BS=0 THEN GOTO 420
390 REM BUT IF EXPLORATION WAS ABOUT A PATTERN PT.
395 REM AND NO REDUCTION WAS MADE CHANGE BASE AT 240
400 REM OTHERWISE REDUCE STEP LENGTH AT 490
410 GOTO 490
420 FOR I=1 TO N:P(I)=B(I):Y(I)=B(I):X(I)=B(I):NEXT I
430 GOSUB 2000:BS=1:PS=0
440 FI=Z:FB=Z
450 PRINT"BASE CHANGE"Z
460 FOR I=1 TO N:PRINT X(I),:NEXT I:PRINT""
470 REM (FOLLOW ON FROM 395)AND EXPLORE ABOUT NEW BASE POINT
480 J=1:GOTO 200
490 K=K/10
500 PRINT"CONTRACT STEP LENGTH"
510 IF K<1E-9 THEN GOTO 700
512 IF R(r)<k THEN GOTO 700
513 R(r+1)=C*R(r)
514 X(I)=X(r)
515 r=r+1
520 REM IF WE HAVE NOT FINISHED MAKE NEW
525 REM EXPLORATION ABOUT LATEST BASE POINT
530 J=1:GOTO 200
535 REM PATTERN MOVE
540 FOR I=1 TO N:P(I)=2*Y(I)-B(I)
550 B(I)=Y(I):X(I)=P(I):Y(I)=X(I)
560 NEXT I
570 GOSUB 2000:FB=FI:PS=1:BS=0:FI=Z
580 PRINT"PATTERN MOVE"Z
590 FOR I=1 TO N:PRINT X(I),:NEXT I:PRINT""
600 REM THEN EXPLORE LATEST PATTERN POINT
610 J=1:GOTO 200
700 PRINT" MINIMUM FOUND"
710 FOR I=1 TO N :PRINT"X"I"="P(I):NEXT I:PRINT""
750 PRINT"FUNCTION MINIMUM"QQ,Q
760 PRINT"NO. OF FUNCTION EVALUATIONS="FE
790 END
2000
Q=0.1334-0.1955*X(2)+1.2729*(X(2)^2)-2.646*(X(2)^3)+378.088
```

```
  *(X(1)^-3)-17.5576*(X(1)^-2)
2004 QQ=R(r)*((1/(X(1)-20))+(1/(80-X(1)))+(1/(X(1)-0))
       +(1/(0.3-X(1))))
2005 Z=Q+QQ
2010 FE=FE+1
2020 REM COUNTS FUNCTION EVALUATIONS
2030 RETURN
```

Appendix C

Re-refining Process Diagrams

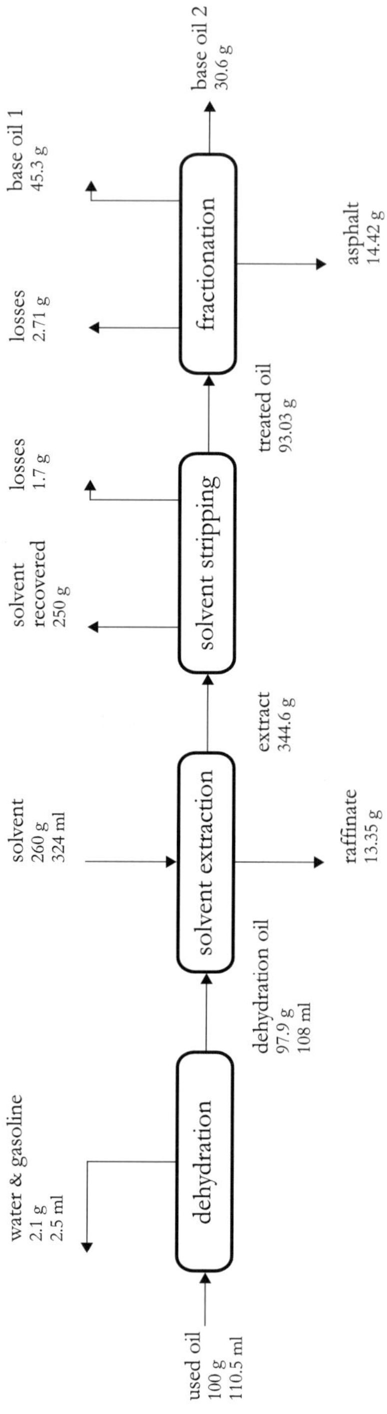

Fig. C.1. Process mass balance.

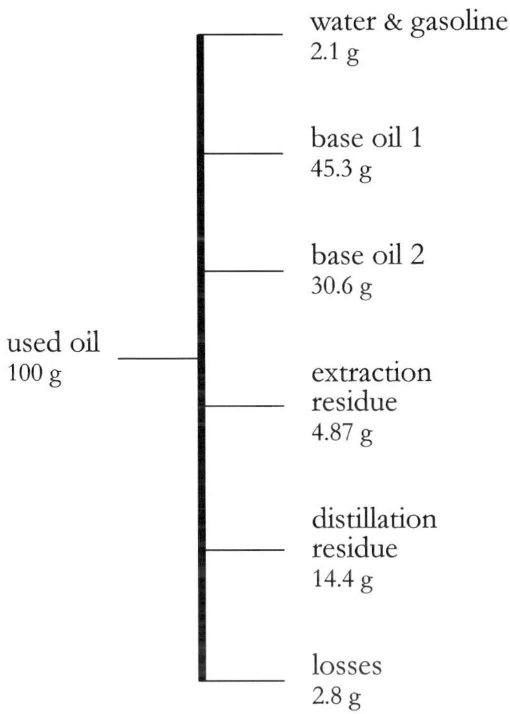

Fig. C.2. Typical process production outcome.

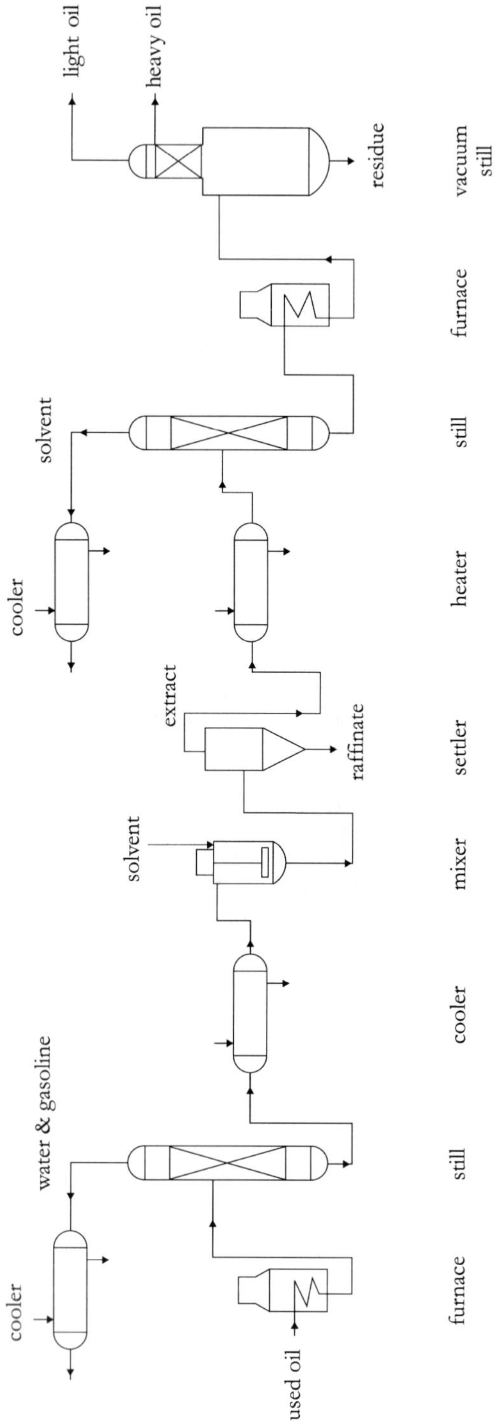

Fig. C.3. Suggested process flow diagram.

References

Adams, M., Awaja, F. and Bhargava, S. (2005). Prediction of oil yield from oil shale minerals using diffuse reflectance infrared Fourier transform spectroscopy, *Fuel* 84 (14–15), 1986–1991.

Ahn, H.S. and Yoon, E.S. (1996). Practical contaminant analysis of lubricating oil in a steam turbine-generator, *Tribology International* 29, 161–168.

Al-Ahmad, M.I. and Al-Mutaz, I.S. (1991). Techno-economic study of re-refining waste lubricating oils in the Arabian Gulf countries, *Resources, Conservation and Recycling* 6, 71–78.

Andreev, V.G. and Tolmachev, G.P. (2002). Future directions in recycling of petroleum-containing wastes. Thermal cracking-optimum method for utilization of waste oils, *Chemistry and Technology of Fuels and Oils* 38 (6).

Arikan, P. and Zararsiz, A. (1996). Determination of wear metals in used lubricating oils by X-ray fluorescence spectrometry, *Nuclear Instruments and Methods in Physics Research, Section B: Beam Interactions with Materials and Atoms* 108, 385–388.

Audibert, F., Defives, D., Avrillon, R. and Minisclaux, C. (1978). Reclaiming of spent lubricating oils by ultra filtration. In: Proceedings of the 3rd International Conference of Used Oil Recovery & Reuse, Texas, USA.

Awaja, F. and Bhargava, S. (submitted). The prediction of clay contents in oil shale using DRIFTS and TGA data combined with multivariate calibration.

Awaja, F. and Pavel, D. (2005). Statistical models for optimization of properties of bottles produced using blends of reactive extruded recycled PET and virgin PET, *European Polymer Journal* 41 (9), 2097–2106.

Barth, E.J. (1962). *Asphalt, Science and Technology*. Gordon and Breach Science Publishers Inc.

Bartlesville (2005). Information found online at http://www.fe.doe.gov/aboutus/history/bartlesville_history_pub.htmL

Bhargava, S., Awaja, F. and Subasinghe, D.N. (2005). Characterization of some Australian oil shale using thermal, X-ray and IR techniques, *Fuel* 84 (6), 707–715.

Bhaskar, T., Azhar Uddin, M., Muto, A., Sakata, Y., Omura, Y., Kimura, K. and Kawakami, Y. (2004). Recycling of waste lubricant oil into chemical feedstock or fuel oil over supported iron oxide catalysts, *Fuel* 83, 9–15.

Berry, R.I. (1981). Re-refining route is set for two plants, *Chemical Engineering* 20, 92.

Bethea, S.R., Bosniack, D.S., Claybaugh, B.E. and Mohundro, E.L. (1973). To hydro treat used lube oil, *Hydrocarbon Processing* 134.

Bishop, J. and Arlidge, D. (1978). Recent technology development in evaporative re-refining of used oil. In: Proceedings of the 3rd International Conference of Used Oil Recovery & Reuse, Texas, USA.

Braithwaite, E.R. (1967). *Lubrication and Lubricants*. Elsevier Publishing Company, London.

Brinkman, D.W. and Dickson, J.R. (1995). Contaminants in used lubricating oils and their fate during distillation/hydrotreatment re-refining, *Environmental Science and Technology* 29, 81–86.

Brown, R.J. (1983). Determination of trace metals in petroleum and petroleum products using an inductively coupled plasma optical emission spectrometer, *Spectrochimica Acta, Part B: Atomic Spectroscopy* 38, 283–289.

Bunday, B.D. (1984), *Basic Optimization Methods*. Arnold, London, UK, Ch. 6.

CEP (2005). Information found online at http://www.evergreenoil.com/cep.htmL

Chen, G. and He, G. (2003). Separation of water and oil from water-in-oil emulsion by freeze/thaw method, *Separation and Purification Technology* 31, 83–89.

Clague, A.D.H. and Donnet, J.B. (1999). A comparison of diesel engine soot with carbon black 1, *Carbon* 37, 1553 1565.

Cochran, W.G. and Cox, G.M. (1975). *Experimental Design*. John Wiley & Sons, NY, USA.

Cotton, F.O., Whisman, M.L., Goetzinger, J.W. and Reynolds, J.W. (1977). Analysis of 30 used motor oil, *Hydrocarbon Processing* 131.

Ekanem, E.J. and Lori, J.A. (1997). The determination of wear metals in used lubricating oils by flame atomic absorption spectrometry using sulphanilic acid as ashing agent, *Talanta* 44, 2103–2108.

Elbashir, N.O. and Al-Zahrani, S.M. (2002). A method of predicting effective solvent extraction parameters for recycling of used lubricating oils, *Chemical Engineering and Processing* 41, 765–769.

Elbashir, N.O., Al-Zahrani, S.M., Abdul Mutalib, M.I. and Abasaeed, A.E. (2002). *Chemical Engineering and Processing* 41, 765–769.

Erdweg, K.J. (1978). Recycling a new process to revert spent oils into lubricant. In: Proceedings of the 3rd International Conference of Used Oil Recovery & Reuse, Texas, USA.

Evergreen (2005). Information found online at http://www.evergreenoil.com/

Forbes, E.S. and Neustadter, E.L. (1972). The mechanism of action of polyisobutenyl succinimide lubricating oil additives, *Tribology* 5 (2), 72–77.

Gary, J.H. and Hardwerk, G.E. (1984). *Petroleum Refining, Technology and Economics*. Marcel Dekker Inc., New York.

Goncalves, I.M. and Murillo, M. (1998). Determination of metals in used lubricating oils by AAS using emulsified samples, *Talanta* 47, 1033–1042.

Gruber, W. (1992). *Evergreen Oils Newark Re-Refining (Running Small, but Making a Big Impact)*. Publication of Environmental Information Ltd.

Gruse, W.A. and Stevens, D.R. (1942). *The Chemical Technology of Petroleum*. McGraw-Hill Book Company, New York.

Guthrie, U.B. (1960). *Petroleum Products Handbook*. McGraw-Hill Book Company, New York.

Hamad, A., Al-Zubaidy, E. and Fayed, M.E. (2005). Used lubricating oil recycling using hydrocarbon solvents, *Journal of Environmental Management* 74 (2), 153–159.

Havemann, R. (1978). The KTI used oil re-refining process. In: Proceedings of the 3rd International Conference of Used Oil Recovery & Reuse, Texas, USA.

Haycock, R.F., Robson, R. and Brice, P. (1988). Understanding black sludge formation, *The Quarterly Review of The Institute of Petroleum* 63.

Hess, L.Y. (1979). *Reprocessing and Disposal of Used Petroleum*. Noyes Data Corporation, NJ, USA.

Hobson, P.D. (1955). *Industrial Lubrication Practice*. Industrial Press.

IFP (2005). Information found online at http://www.ifp.fr/IFP/en/aa.htm

Kalichevsky, U.A. and Kobe, K.A. (1956). *Petroleum Refining with Chemicals*. Elsevier Scientific Publishing Company, London.

Kim, S.S., Chun, B.H. and Kim, S.H. (2003). Nonisothermal pyrolysis of waste automobile lubricating oil in a stirred batch reactor, *Chemical Engineering Journal* 93, 225–231.

Krell, E. (1982). *Handbook of Laboratory Distillation*. Elsevier Scientific Publishing Company, London.

KTI (2005). Information found online at http://www.specialized-mechanical.com/kti.htmL

Martins, J.P. (1997). The extraction–flocculation re-refining lubricating oil process using ternary organic solvents, *Ind. Eng. Chem. Res.* 36, 3854–3858.

McKeagan, D.J. (1992). Economics of re-refining used lubricants, *Lubrication Engineering* May.

McKetta, J.J. (1986). *Encyclopaedia of Chemical Processing & Design* 28, 378–417.

Miller, R.E. (1986). Experimental design, *Chemical Engineering* 113.

Mohammed, A.A., Hankish, K., Nagi, A.R., Abbas, N. and Said, M.A. (1984). Re-refining and uses of spent lubricating oil, *Journal of Petroleum Research* 1 (3), 15.

Monick, J.A. (1968). *Alcohols, Their Chemistry, Properties and Manufacture*. Reinhold Book Corporation, New York.

Nelson, W.L. (1964). *Petroleum Refining Engineering*, 4th edn. McGraw-Hill International Book Co.

Omar, A.M.A., El Adly, R.A., Keera, S.T. and Mohamed, M.S. (1998). Recovery of residual lubricating oil from waste clay by flotation, *Monatshefte für Chemie* 129, 387–392.

Ostrikov, V.V. and Prokhorenkov, V.D. (2003). Waste-free technology for processing used lubricating oil, *Chemical and Petroleum Engineering* 39, 5–6.

Patel, A.J. (1994). Liquid used disposal and recovery-lubricant recycling, *Industrial Lubrication & Tribology* July/Aug.

Pauley Jr., J.F. (1978). Thin-film distillation as a tool in re-refining used oil. In: Proceedings of the 3rd International Conference of Used Oil Recovery & Reuse, Texas, USA.

Quang, V.D., Bonneford, P. and Kerdraon, L. (1976). Spent oil reclaimed without acid, *Hydrocarbon Processing* December, 130–131.

Rao, S.S. (1979). *Optimization, Theory and Application*. Wiley Eastern Limited, New Delhi.

Reis, M.A. and Jeronimo, M.S. (1988). Used lubricating oil re-refining by extraction–flocculation 1-A scientific basis to design efficient solvents, *I&EC Res.* 27, 1222–1228.

Reis, M.A. and Jeronimo, M.S. (1990). Used lubricating oil re-refining by extraction–flocculation 2-A method to formulate efficient composite solvent, *I&EC Res.* 29, 432–436.

Reynolds, J.W., Whisman, M.L., Brinkman, D.W., Goetzinger, J.W. and Cotton, F.O. (1979). From oil: oil, *Chemtech* 628.

Robinson, C.S. and Gilliland, E.R. (1950). *Elements of Fractional Distillation*, 4th edn. McGraw-Hill Book Company Inc., NY, USA.

Salvador, A. and Guardia, M. (1983). Determination of the total iron content of used lubricating oils by atomic-absorption with use of emulsions, *Talanta* 30, 986–988.

Scott, W. and Hargreaves, D. (1991). A case for reclaimed lubricating oil. In: Int. Mech. Eng. Congress & Exhibition MECH.-91, Sydney, Australia.

Simiele, G.A. and Field, R.A. (1987). Radioactive decontamination of waste oil by filtration, centrifugation, and chelation, *Nuclear and Chemical Waste Management* 7, 257–263.

Skala, D.U., Saban, D.K., Orlovic, A.M., Meyn, V.W., Severin, D.K., Rahimian, G.H. and Marjanovic, M.V. (1991). Hydrotreating of used oil: prediction of industrial trickle-bed operation from pilot plant data, *Industrial and Engineering Chemical Research* 30, 2059–2065.

Snow, R.J. and Delaney, S.F. (1977). Vacuum distillation of used lubricating oil, *Chemeca* 77, 14–16.

Van Dyk, J.W., Frisch, H.L. and Wu, D.T. (1985). Solubility. Solvency and solubility parameters, *I&EC Prod. Res. Dev.* 24, 473–478.

Viscolube (2005). Information found online at http://www.viscolube.it

Wang, J., Jia, C.R., Wong, C.K. and Wong, P.K. (2000). Characterization of polycyclic aromatic hydrocarbons created in lubricating oils, *Water, Air, and Soil Pollution* 120, 381–396.

Whisman, M.L., Goetzinger, J.W. and Cotton, F.O. (1974). Used Lubricating Oil Research. Some Innovative Approach Reclaiming Used Crankcase Oil. USA 234685.

Whisman, M.L., Reynolds, J.W., Goetzinger, J.E. and Cotton, F.O. (1978). Re-refining makes quality oils, *Hydrocarbon Processing* October, 141–145.

Subject Index